The A-Z of Astronomy

Patrick Moore was born in 1923, and read his first
book on astronomy six years later. When still of
school age he was elected to the British Astronomical
Association, and concentrated on observations of the
Moon. During the war he flew as a navigator in the
Royal Air Force; subsequently he set up his private
observatory in Sussex and now lives in Selsey. Between
1965 and 1968 he was Director of the Armagh
Planetarium in Northern Ireland.

Every month since April 1957 Patrick Moore has
presented the BBC Television programme *The Sky at
Night*, and he is also a frequent broadcaster on radio.
He is Director of the British Astronomical Association
Lunar Section and the Association's Goodacre
Medallist. His hobbies include cricket, music and
chess. He was awarded the OBE in 1968.

D1440226

THE A-Z OF
ASTRONOMY

Patrick Moore

Diagrams by Cyril Deakins

FONTANA/COLLINS

First published as *The Amateur Astronomer's Glossary*
by the Lutterworth Press 1966
This revised edition first published by Fontana 1976

Copyright © Patrick Moore 1966, 1976

Made and printed in Great Britain
by William Collins Sons & Co. Ltd Glasgow

CONDITIONS OF SALE
This book is sold subject to the condition that
it shall not by way of trade or otherwise. be lent,
re-sold, hired out or otherwise circulated without
the publisher's prior consent in any form of
binding or cover other than that in which it is
published and without a similar condition
including this condition being imposed on the
subsequent purchaser.

Foreword

A complete dictionary of astronomy would be a very large volume indeed. In this brief 'A–Z', I have made no attempt to be complete, and I am well aware that many terms of importance have been left out; however, I hope that the result will be useful for quick reference.

In the text, an asterisk indicates that the term so marked is described under its own separate heading.

The Planetarium,
Armagh, 1966

Foreword to the Revised Edition

Much has happened in the world of astronomy since this book was first published, so I have taken the opportunity to bring the text up to date. I am most grateful to Michael Foxall and to Hilary Rubinstein for all their help and encouragement.

Selsey, 1975   PATRICK MOORE

A

Aberration of Starlight. The apparent displacement of a star from its true position in the sky, due to the fact that light has a definite velocity (186,000 miles per second) and does not move infinitely fast. A good analogy is to picture a man walking along in a rainstorm, holding up an umbrella to shield himself. To keep dry, he will have to slant the umbrella forward, as shown in the diagram (Fig. 1); in other words, the raindrops will seem to be coming at an angle, instead of straight down. In the case of starlight, the aberration effect is due to the movement of the Earth, which is travelling round the Sun at an average velocity of 18½ miles per second; thus the starlight seems to come 'at an angle'. The apparent positions of stars may be affected by up to 20·5 *seconds of arc.

Fig. 1. *Aberration*

Absolute Magnitude. The *magnitude that a star would seem to have if it were observed from a distance of 10 *parsecs, or 32·6 *light-years. (See *Magnitude.*)

Absorption of Light in Space. It was formerly thought that space must be completely empty. This is now known to be wrong; there is a vast amount of thinly-spread interstellar matter, so that the light coming from distant objects is partially absorbed.

Achromatic Lens. A lens corrected for chromatic aberration, so that 'false colour' is reduced.

Adams, John Couch (1819–1892). Great English mathematical astronomer, chiefly remembered for his correct prediction of the position of Neptune – though the actual discovery of Neptune was made on the basis of similar calculations by the French astronomer *Le Verrier.

Adams, Walter S. (1876–1936). American astronomer; Director of *Mount Wilson Observatory from 1923 to 1946. His main work was in stellar spectroscopy.

Aerolite. A stony *meteorite.

Airglow. The faint luminosity of the night sky, due mainly to processes going on in the Earth's upper atmosphere.

Airy Disk. The apparent size of a star-disk produced by a perfect optical system – since the star can never be focused perfectly, 84% of the light will concentrate into a single disk, and 16% into a system of surrounding rings.

Airy, Sir George Biddell (1801–1892). The seventh Astronomer Royal, who was appointed to the post in 1835 and retired in 1881. He was a great administrator, though somewhat autocratic! He made many notable contributions to astronomy, and was responsible for raising Greenwich Observatory to its present position of eminence.

Aitken, Robert (1864–1949). A great American observer of double stars, and one-time Director of the *Lick Observatory.

Albedo. The reflecting power of a planet or other non-luminous body; the ratio of the amount of light reflected from the body, to the amount of light which falls on to the body from an outside source. A perfect reflector would have an albedo of 100%.

Alcyone. The brightest star in the *Pleiades.

Algol. The prototype *eclipsing binary.

Alpha Centauri. The nearest of the bright *stars; distance 4·3 light-years. It is a fine binary, but too far south to be seen from

Europe. Its faint companion, Proxima, is slightly closer to us (4·2 light-years).

Altazimuth Mount. A type of telescope mount upon which the instrument may swing freely in any direction (Fig. 2). Small telescopes are often mounted in this way, but a larger telescope is better fitted with an *equatorial mount.

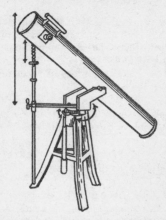

Fig. 2. *Altazimuth mount*

Altitude. The angular distance of a celestial body above the horizon. At the overhead point or *zenith, the altitude is 90°; at the horizon, the altitude is 0°.

Amalthea. The innermost satellite of Jupiter; it is the closest to the planet of the Jovian satellite family, but has a diameter of only about 150 miles. It was discovered by Barnard in 1892.

Ananke. The twelfth satellite of Jupiter.

Andromeda Galaxy. One of the nearest of the external star-systems or *galaxies. It is dimly visible to the naked eye; photographs with large telescopes show it to be spiral in structure. It lies at a distance of 2·2 million *light-years, and is considerably larger than the Galaxy in which we are situated. The galaxy is officially known as M.31, since it was the 31st object in a famous catalogue drawn up by the French astronomer Charles Messier in 1781.

Ångström Unit. The unit for measuring the wavelength of light, X-rays and other electromagnetic vibrations. It is equal to one hundred-millionth part of a centimetre. The wavelength of visible light ranges between about 3900 Ångströms (violet) and 7500 Ångströms (red). The unit is named after the last-century Swedish physicist Anders Ångström; the official abbreviation is Å, though often written simply as A.

Antoniadi, Eugene M. (1870–1944). Greek-born astronomer who spent most of his life in France, and became the most distinguished planetary observer of his time. He was particularly noted for his studies of Mercury and Mars, carried out mainly with the 33-inch refractor at the Observatory of *Meudon.

Antares. The leading *star in Scorpio; a Red Giant over 250,000,000 miles in diameter. Its name means 'the Rival of Mars'. It has a small greenish companion which is a radio source.

Apennines. The most conspicuous mountain range on the *Moon. It borders the Mare *Imbrium.

Aphelion. The position in the *orbit of a planet or other body when furthest from the Sun. For instance, the Earth is at its furthest point, or aphelion, in early July, when its distance from the Sun is 94½ million miles; at its closest to the Sun, in early January (*perihelion) the distance is only 91½ million miles. Similarly, *apogee* refers to a body moving round the Earth; the Moon is at apogee when at its greatest distance from us.

Areography. The physical study of *Mars (from Greek, *Ares*).

Ariel. The first *satellite of Uranus.

Aristarchus (BC 310–250). Born at Samos. Greek astronomer, who believed the Earth to be in motion round the Sun, and who made a noble attempt to measure the distances of the Moon and Sun.

Armagh Observatory. The leading observatory in Northern Ireland, founded in 1790.

Armillary Sphere. An old type of instrument which may be described as a 'skeleton' of the celestial sphere.

10

Artificial Satellite. A man-made vehicle put into a path or *orbit round the Earth. It is launched by rocket power; once it has been put into a stable orbit it will not come down unless it spends any part of its time within the resisting part of the *atmosphere, in which case it will be affected by friction against the air-particles and will gradually have its orbit changed until it re-enters the lower atmosphere and is destroyed. Of course, an artificial satellite may be brought down deliberately in a con-trolled descent, as has happened with many of the Russian vehicles – and, needless to say, with all man-carrying satellites!

The first artificial satellite, Sputnik I, was launched by the Russians on 1957 October 4: this may be said to mark the real beginning of the Space Age. Many more have followed, most of which have been launched by the Americans or the Russians, though a few have come from Britain, Italy, China, Japan and France. The first US satellite, Explorer I of 1958, was of special importance, since its instruments detected the zones of intense radiation round the Earth now known as the *Van Allen zones.

Artificial satellites have been put to many uses. They have been invaluable sources of information about Earth's resources, and they are also used daily as communications relays for both radio and television. Instruments carried in satellites have studied phenomena such as *cosmic rays, *meteoroids, *ultra-violet and *X-radiations from space, and so on. Meteorological satellites have led to a great advance in our knowledge of the weather systems, and have also done great service in giving advance warnings of potentially dangerous storms developing out at sea. Many lives have been saved in this way.

Some of the artificial satellites have been very bright – notably the two Echo balloon vehicles of the 1960s. Most, however, are considerably fainter, and look like slowly moving stars.

Ascending Node. This is described under the heading *Nodes·

Ashen Light. When the planet *Venus appears as a crescent, the 'night' side sometimes appears dimly luminous. This is known as the Ashen Light. Its cause is not certainly known; some author-ities believe it to be a sheer contrast effect, but more probably it is a genuine phenomenon. There is no comparison with the *earthshine seen on our Moon, since Venus has no satellite.

Asteroids. An alternative name for the *Minor Planets.

11

Astræa. The fifth *asteroid, discovered by Hencke in 1845. It is considerably smaller than the first four members of the swarm.

Astrograph. A telescope designed specifically for astronomical photography.

Astrolabe. An instrument used by ancient astronomers to measure the *altitudes of bodies in the sky. An astrolabe consisted of a circular disk marked off in *degrees along its rim; the star or planet was sighted by means of a movable arm, with the astrolabe held suspended in a vertical position, and the altitude read off upon the scale.

Astrology. The so-called 'science' which attempts to link human destinies and characters with the positions of the planets against the starry background. It is still widely practised in some countries, notably India; but it has no scientific foundation whatsoever, and has long since been completely discredited. It should never be confused with true *astronomy.

Astronautics. The science of space research, using either unmanned vehicles or manned space-ships. It is a modern development, since it is only since the end of the war that rockets have become powerful scientific tools instead of mere toys; but since 1957 it has become of fundamental importance, and has provided information which could not possibly have been obtained in any other way.

Astronomical Unit. The distance between the Earth and the Sun; in round figures, 93,000,000 miles. The mean distance from the Earth to the Sun has now been found to be slightly less than this (92,957,209 miles or 149,598,500 km) but the astronomical unit is still conventionally kept as 93,000,000 miles.

Astronomy. The science dealing with the bodies in the sky. It began in prehistoric times, when our remote ancestors gazed up at the stars and formed them into fanciful *constellations; but its real development began with the Greeks, who drew up excellent star catalogues, measured the size of the Earth, and studied the movements of the Sun, Moon and planets. Unfortunately, most of them made the mistake of supposing that the Earth must lie in the centre of the whole universe.

*Ptolemy, last of the great astronomers of ancient times, died about AD 180, and for some centuries after this little progress

12

was made. A revival came with the Arabs of a thousand years ago, and at last interest in astronomy was re-kindled in Europe. In 1546 Copernicus, a Polish churchman, published a book in which he rejected the Greek theories of the universe, and claimed that the Earth is nothing more than a planet moving round the Sun. Arguments about this vitally important problem went on for more than a century, but the work of great scientists such as Galileo and Kepler, followed by Sir Isaac Newton's researches into the effects of *gravitation, finally settled the matter in favour of Copernicus.

Telescopes were invented in the early 17th century, and were used astronomically by Galileo in the winter of 1609–10. Though Galileo was probably not the first telescopic astronomer, he was certainly the most skilful, and he made a series of spectacular discoveries; for instance he observed the satellites of Jupiter, the phases of Venus and the countless stars in the Milky Way. With the construction of more powerful telescopes, progress in astronomy became rapid, and during the 18th and 19th centuries came the development of astronomical *photography and the *spectroscope. More recent still are *radio astronomy and *rocket astronomy.

Modern astronomy is, of course, basically mathematical, and the professional astronomer does very little direct observation at the eye-end of a telescope. Amateurs still carry out useful work with relatively modest equipment, but any student wishing to make a career as a professional astronomer must make sure that he (or she) is a first-class mathematician.

Astrophysics. The branch of modern astronomy which may be defined as 'the physics and chemistry of the stars'. Further details will be found under the headings *Stars* and *Spectroscope*.

Atmosphere. The gaseous mantle surrounding a planet or other body. The Earth's atmosphere is made up of several layers; the bottom seven miles or so is known as the *troposphere, above which come more rarefied layers such as the *stratosphere, *ionosphere and finally the *exosphere, which has no hard and sharp boundary, but which simply 'tails off' into space (Fig. 3). The atmosphere may be said to extend upward for at least 2000 miles above the ground, though most of the mass of the atmosphere is concentrated in the bottom four or five miles.

The Earth, with an *escape velocity of 7 miles per second, has been able to retain a dense atmosphere. Bodies with lower escape velocities, such as Mars (3 miles per second) have thinner atmospheres, while the Moon (escape velocity $1\frac{1}{2}$ miles per second) has virtually no atmosphere at all. On the other hand, the giant planet Jupiter (escape velocity 37 miles per second) has been able to hold down even hydrogen, lightest of all the gases, so that its atmosphere is made up largely of hydrogen and hydrogen compounds.

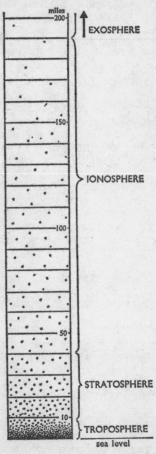

Fig. 3. *Cross-section of the atmosphere*

Atom. The smallest unit of a chemical *element which retains its own particular character. Each atom may be said to consist of a central nucleus, around which move particles known as *electrons; the nucleus has a positive electrical charge and the electrons have negative charges, the two balancing each other and making the complete atom electrically neutral. Modern atomic theory is highly complicated, and it is quite misleading to suppose that an atom is a miniature Solar System, made up of solid lumps, but it is almost impossible to give a useful description in non-technical language.

Atoms may link up to form atom-groups or *molecules*; thus a molecule of water is made up of two hydrogen atoms combined with one oxygen atom, so that its chemical formula is the familiar H_2O.

Aurora. Auroræ are the lovely Northern Lights (Aurora Borealis) and Southern Lights (Aurora Australis). They occur in the Earth's upper atmosphere, and are due to charged particles sent out by the Sun, which penetrate the outer air and produce the beautiful glows. The process is not yet completely understood, and is connected with the zones of trapped particles in the *Van Allen Zones. At any rate, auroræ are linked with events taking place in the Sun; a brilliant solar *flare, which emits charged particles, is likely to be followed by a display of aurora about 24 hours later.

Because the particles are magnetic, they tend to move toward the Earth's magnetic poles. Consequently, auroræ are best seen in the polar regions. In North Norway, Iceland and even North Scotland, auroræ are common during the winter; in South England, brilliant displays are rare, and in low latitudes auroræ are almost (though not quite) unknown. Obviously, auroræ are commonest when the Sun is at its most active. The next solar maximum is not due until about 1980.

Azimuth. The angular bearing of an object in the sky, measured from north (0°) through east (90°), south (180°) and west (270°) back to north (360° or 0°). Due to the Earth's rotation, the azimuths and *altitudes of all celestial bodies change constantly.

B

Baade, Walter (1893–1960). A German-born astronomer who spent most of his life in the United States. He made many notable contributions to stellar astronomy, and discovered the error in the *Cepheid scale which led to a doubling of the estimated distances of the *galaxies.

Baily's Beads. Brilliant points seen along the edge of the Moon's dark disk at a total solar *eclipse, just before and just after actual totality. They are caused by the Sun's light shining through valleys on the limb of the Moon, between mountainous regions.

Baily, Francis (1774–1844). English amateur astronomer, best remembered for his observations of *Baily's Beads at the total solar eclipses of 1836 and 1842.

Balloon Astronomy. From the Earth's surface, some types of astronomical observations cannot be carried out, because we are looking at the sky through a dense layer of atmosphere. For instance, it proved to be impossible to detect water-vapour in the atmosphere of the planets Mars or Venus, because the light from these bodies had to pass through our own atmosphere – which contains considerable water-vapour, and masks the traces of water-vapour in the atmospheres of other worlds.

One solution is to send up instruments in balloons to heights of 85,000 feet or so, above most of the Earth's air. Several very successful ascents have been made during the last few years, and there can be no doubt that balloon astronomy has an important future. Though it is naturally limited in scope, it has two great advantages over *rocket astronomy: balloons are easy to control, and are also easy to recover intact, together with their instruments and all their scientific records. Moreover, they are cheap!

Barnard, Edward Emerson (1857–1923). American astronomer, noted for his observations of planets and for his comet discoveries. In 1892 he discovered *Amalthea, the fifth satellite of Jupiter.

Barycentre. The centre of gravity of the Earth-Moon system. The Earth is 81 times more massive than the Moon, so that the barycentre lies inside the Earth's globe (Fig. 4).

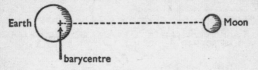

Fig. 4. *The barycentre*

Beer, Wilhelm (1797–1850). German amateur astronomer, who collaborated with J. H. *Madler in compiling the first good map of the Moon.

Bessel, Friedrich (1784–1846). Great German astronomer, who made many contributions but is best remembered as being the first man to announce a measurement of the distance of a star (61 Cygni, in 1838).

Betelgeux. The second brightest *star in Orion; a Red Giant, slightly variable in brilliancy, and extremely large, with a diameter of over 250,000,000 miles.

Biela's Comet. A comet which used to move round the Sun in a period of 6¾ years. At the return of 1845 it split in two; the 'twins' were seen for the last time in 1852. In 1872 a meteor shower was seen instead of the comet. There is no doubt that Biela's Comet is now defunct.

Binary Star. A star made up of two components, genuinely associated, and moving round their common centre of gravity in rather the fashion of two dumb-bells when twisted by their joining bar (Fig. 5).

Binaries are very common in space. In some cases the components are equal; such is Theta Serpentis, which may be separated with a small telescope. With other pairs, the components may be somewhat unequal, as with Mizar or Zeta Ursæ Majoris, the second star in the tail of the Great Bear, where one member of the pair is obviously brighter than the other. There are also many binaries in which one component far outshines the other; thus the brilliant Sirius is accompanied by a faint star of the type known as a *White Dwarf.

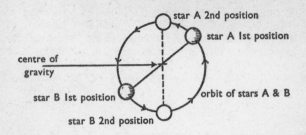

Fig. 5. *Movements of binary components*

The periods of revolution have a considerable range. With pairs which are widely separated, the period may amount to tens, hundreds or many thousands of years, but close pairs have short periods. When the separation is very small indeed, the components cannot be seen individually, though they may be detected by other methods; pairs of this kind are known as *spectroscopic binaries*.

There are also many cases of 'stellar families', or multiple stars. Particularly noteworthy is Castor, in the constellation of Gemini (the Twins). A moderate telescope will show that Castor is made up of two; each component is again double, and there is a third, fainter companion which is itself a close binary. Altogether, Castor consists of six suns, four bright and two feeble.

It used to be thought that a binary star must be the result of the break-up or fission of a formerly single star, but it is now regarded as more probable that the components were always separate from each other so that presumably they were formed at the same time in the same region of space.

Black Hole. A localized region of space from which not even light can escape, because of the presence of a small, super-dense collapsed star or *collapsar.

So far as we know, a star with a mass of more than about ten times that of the Sun cannot become a White Dwarf or an ordinary *neutron star. When it starts to collapse, after its nuclear energy is used up, gravitation takes over, and the material is crushed together until a completely new situation arises. If the *escape velocity becomes greater than the velocity of light (186,000 miles per second) the star will be unobservable. It will be surrounded by a kind of 'forbidden zone' making up the black hole; the boundary of this zone is termed the event horizon. Inside this event horizon, all the ordinary laws of physics break down. It has even been suggested that the collapsed star may be crushed out of existence altogether, but the black hole will still exert a gravitational pull on other bodies, so that it will be detectable even though it cannot be seen.

Obviously, we can track down black holes only if they are reasonably close to visible stars – in fact, if they are members of *binary systems. One possible case is Epsilon Aurigæ, which lies near Capella in our sky (though, of course, it is far more remote, and there is no connection between the two). Epsilon Aurigæ has a primary component which is a very powerful supergiant, and is easily visible with the naked eye. Every 27 years the primary is eclipsed by an invisible secondary, and the total brightness of the system drops by over a magnitude. The secondary may be a very young star still contracting toward the *Main Sequence, but there is also a suggestion that it may be a black hole. It is an infra-red source, and on the latter interpretation this radiation is due to a cloud of solid particles moving round the collapsar at a distance of around 15,000 million miles from the centre of the black hole.

Even more significant is Cygnus X-1, which is an X-ray source. Here the visible component is another supergiant, known by its catalogue number of HDE 226868, with a mass 30 times that of the Sun. The invisible secondary has a mass of about half this value – and if it were an ordinary star, it should certainly be visible.

It must be stressed that we have no positive proof that black holes exist, and alternative theories have been put forward to explain the peculiarities of Cygnus X-1. Possibly we are dealing with two stars whose very strong magnetic fields become entwined as the stars spin round; the magnetic lines of force are periodically broken and then re-connect, with the release of energy and the production of X-rays. Moreover, some very bizarre ideas have been put forward – such as the suggestion that a black hole could abruptly reappear in another area as a 'white hole'! This sounds remarkably like science fiction. Yet the whole concept of black holes is extraordinary by any standards, and we can only await the results of future research.

Blink-Microscope (or *Blink Comparator*). An instrument for examining two photographs in rapid succession. If the same starfield is photographed at two different times, and the pictures are shown alternatively in rapid succession – three or four 'blinks' per second – a moving object, such as a *minor planet, will seem to jump to and fro, thereby making itself conspicuous. Various bodies, including the planet *Pluto, have been first identified in this way. Blink-microscopes are also used for studies of the *proper motions of the stars.

Bliss, Nathaniel (1700–1764). The fourth Astronomer Royal (1762–4).

BL Lacertæ Objects. Variable objects in both the optical and radio range; they look superficially stellar, but near minimum light show a fuzzy edge. They are very powerful emitters of infra-red radiation, which implies the presence of dust. They are extremely remote, and may be intermediate in luminosity between *quasars and *Seyfert galaxies; their nature is unknown, but it has been suggested that they are immature quasars. BL Lacertæ itself was identified by M. Schmidt in 1968; previously it had been thought to be an ordinary variable star, but Schmidt found that the spectrum is absolutely featureless. It may rise to magnitude 12, and is the brightest member of the class; others include AP Libræ and W Comæ.

Bode's Law. An interesting relationship between the distances of the planets from the Sun, first noticed by J. D. Titius in 1772, but made famous by J. E. Bode. It may be summed up as follows:

Take the numbers 0, 3, 6, 12, 24, 48, 96, 192 and 384, each of which (apart from the first) is double its predecessor. Add 4 to each. Taking the Earth's distance from the Sun as 10, the distances of the other planets known in Bode's time are given with remarkable correctness, as follows:

| | | *Distance:* |
Planet	Bode's Law	Actual
Mercury	4	3·9
Venus	7	7·2
Earth	10	10
Mars	16	15·2
Jupiter	52	52·0
Saturn	100	95·4

Uranus, discovered in 1781, was found to have a distance of 191·8; Bode's Law had predicted 196, while the gap corresponding to the missing number 28 was filled by the *minor planets, of which the first, Ceres, was discovered in 1801 and fitted excellently into the scheme. However, Neptune, discovered in 1846, departs from the rule, since its distance is only 300·7 instead of the predicted 388; the last number corresponds much more closely to Pluto (distance 394·6), which was discovered only in 1930.

It is not known whether Bode's Law is of real significance, or whether it is due to pure chance. It does not apply to the satellites of the major planets.†

Bolide. A brilliant *meteor, which may explode during its descent through the Earth's atmosphere.

Bolometer. A very sensitive electrical instrument, used to measure slight quantities of heat radiation.

Boyden Observatory. A leading South African observatory, at Bloemfontein; the present Director is Dr A. H. Jarrett. Its largest telescope is a 60-inch reflector.

Bradley, James (1693–1762). The third Astronomer Royal (1742–62). He discovered the *aberration of light, and compiled a very accurate star catalogue.

Brahe, Tycho (1546–1601). Danish astronomer, whose accurate measurements of star positions and the movements of Mars led to the discovery, by *Kepler, that the Sun rather than the Earth is the centre of the Solar System – something which Tycho himself could never accept. Tycho was the last great astronomer of pre-telescopic times.

† Bode's Law has no connection with the modern, totally unofficial 'Spode's Law', which states, broadly, that 'if things *can* go wrong, they *do*'!

C

Calendar. The system of dividing time into convenient periods – such as days, weeks and months – for our everyday needs. The obvious basic unit is the time taken for the Earth to go once round the Sun, generally termed the 'year', but this raises difficulties, because the true 'year' is not an exact number of days. While completing a full journey round the Sun, the Earth spins on its axis 365¼ times, but to have a calendar taking quarter-days into account would be absurd. Therefore, we take the year as being 365 days, and add on an extra day every fourth year to make up for the difference; thus every fourth year is a Leap Year, with the shortest month, February, having 29 days instead of its usual 28.

An easy way to tell which years are Leap Years is to divide by 4. If there is no remainder, then the year is a Leap Year. The only exceptions are the 'century years' (1700, 1800, 1900 etc.) in which the division is by 400. Thus 1900 was not a Leap Year, but 2000 will be.

Various calendars have been used in the past. Our own is adapted from that of the Romans. Originally there were only ten months, the first of which was March; the great dictator Julius Caesar introduced the 'Julian Calendar', which was first used in BC 44 and was a notable improvement; a further modification was made in 1582 by order of Pope Gregory XIII. Britain adopted this *Gregorian Calendar* in 1752.

Callisto. The fourth Galilean satellite of Jupiter; it is considerably larger than the Moon, and comparable with Mercury, but it has a relatively low *albedo. See *Satellites.

Caloris Basin. A large depression on *Mercury, photographed from Mariner 10 in 1974.

Campbell, W. W. (1862–1938). American astronomer; Director of the Lick Observatory, 1901–30. He was a pioneer in studies of the *radial velocities of stars.

Cannon, Annie J. (1863–1941). American woman astronomer, who worked at the Harvard College Observatory and was concerned mainly with stellar spectra.

22

Canopus. The second brightest of the fixed *stars; Alpha Carinae. It is very luminous and remote. It is too far south to be seen from Europe.

Cape Observatory. Leading South African Observatory, and still the administrative centre for the Republic, though the main telescope – a 40-inch reflector – has now been moved to *Sutherland.

Carbon-Nitrogen Cycle. The stars are not 'burning' in the usual sense of the word; they are producing their energy in a more complex fashion. Most stars contain a great deal of the light gas hydrogen, and this hydrogen is being steadily changed into another gas, helium, with release of energy and loss of mass. One way in which this conversion of hydrogen into helium takes place is by a whole series of reactions, in which two more elements, carbon and nitrogen, are concerned; this is the carbon-nitrogen cycle. It was once thought that the Sun shone because of this process, but modern work has shown that another cycle, the so-called *proton-proton reaction, is more important in stars of solar type. However, the end result is precisely the same; hydrogen turns into helium, and the star continues to shine.

Carbon Stars. Some red stars of spectral types R and N, containing an unusual amount of the element carbon in their atmospheres. (See *Spectroscope*.)

Carme. The eleventh satellite of Jupiter.

Carrington, Richard (1826–1875). English amateur astronomer; a pioneer in observations of the *Sun.

Cassegrain Reflector. A type of reflecting telescope in which the main mirror has a hole in the middle (Fig. 6). The light from the object to be observed passes down the telescope tube and is reflected from the main mirror on to a smaller, convex mirror termed the *secondary*. The light is then sent back through the hole, and the image produced is magnified by an *eyepiece in the usual way. With a Cassegrain, therefore, the observer looks up the instrument instead of into the tube, as with a *Newtonian reflector.

The Cassegrain system has many advantages, and is becoming more and more popular, but although the tube will be shorter than that of an equal-aperture Newtonian the telescope is less easy to construct.

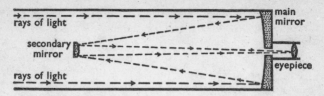

Fig. 6. *Principle of the Cassegrain reflector*

Cassini Division. The principal division in *Saturn's ring-system.

Cassini, Giovanni (often known as Jean Domonique Cassini; although Italian-born he spent much of his life in France). (1625–1712). First Director of the Paris Observatory. He was a great observer of the planets, and discovered four satellites of Saturn, as well as the main division in the planet's rings.

Celestial Sphere. An imaginary sphere surrounding the Earth (Fig. 7). For this purpose we may suppose that the sky is solid (as the ancients used to believe!), and that we lie in the exact centre of the sky-sphere. This is shown in the diagram, with the Earth in the middle and the celestial sphere beyond. A useful, though imperfect, way to show what is meant is to picture a table-tennis ball suspended in the exact centre of a football; the table-tennis ball then stands for the Earth, while the inside surface of the football represents the celestial sphere.

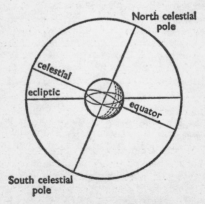

Fig. 7. *The celestial sphere*

The Earth spins on its axis once a *day (roughly 24 hours). The axis points northward to the *north celestial pole*, or pole of the sky, marked approximately by the bright star Polaris in the constellation of Ursa Minor, the Little Bear; there is no bright star at the *south celestial pole*. As the Earth rotates, the celestial poles seem to remain stationary, while the rest of the sky moves round. From northern latitudes, the south celestial pole can never be seen, as it always stays below the horizon; it follows, of course, that people who live south of the equator can never see Polaris. The celestial sphere is divided into two equal hemispheres by the celestial *equator.

Cepheid. An important type of *variable star. Cepheids have short periods of from a few days to a few weeks; they are perfectly regular, and it has been found that the real luminosity of a Cepheid is linked with its period of variation. The longer the period (that is to say, the interval between one maximum and the next), the more luminous the star; thus a Cepheid with a period of 7 days will be more powerful than a Cepheid whose period is only 6 days. From this, it follows that once a Cepheid has been studied, and its period measured, its real luminosity can be found, which in turn allows its distance to be calculated. Of course, many corrections have to be made, notably for the *absorption of light in space, but the basic principle is clear enough.

Cepheids are brilliant stars, far more luminous than our Sun, so that they are visible across great distances, and provide astronomers with a great deal of information; they have been nicknamed our 'standard candles'. One example may help to show what is meant. In 1923, the American astronomer E. E. Hubble detected Cepheids in the *Andromeda Galaxy. At the time it was still not known whether the Andromeda Galaxy lay in our own star-system, or whether it were a separate galaxy at a much greater distance. As soon as Hubble measured the periods of the Cepheids, and obtained their distances, he saw that they were so remote that they must lie far beyond the limits of our Galaxy. This meant that the Andromeda object could only be an independent system.

It has since been found that there are two distinct classes of Cepheids, but the general principle of the *period-luminosity law* is undoubtedly valid, even though we are still uncertain about its cause. The name 'Cepheid' has been given because the star Delta Cephei, easily visible with the naked eye, is the most conspicuous member of the class.

Ceres. The largest and first-discovered of the *Minor Planets. It was found by the Italian astronomer Piazzi in 1801, and has a diameter of about 700 miles (Fig. 8). Its orbit lies between those of Mars and Jupiter, and its distance corresponds well to the 'missing figure' 28 of *Bode's Law. Ceres never becomes bright enough to be visible with the naked eye.

Fig. 8. *Sizes of Ceres and Britain compared*

Christie, Sir William (1845–1922). The eighth Astronomer Royal (1881–1910); like his predecessor, *Airy, a great administrator.

Chromatic Aberration. A defect found in all lenses, resulting in the production of 'false colour'. Light is a mixture of various wavelengths; for visible light, violet has the shortest wavelength (3900 Ångströms) and red the longest (7500 Ångströms). When passed through a lens, the shorter wavelengths will be the more bent or *refracted*, so that violet will be brought to focus closest to the lens (Fig. 9). In fact, the various colours are brought to focus in different places, so producing a whole series of images instead of only one; the result is an annoying amount of false colour.

By using a compound lens, made up of several parts of different kinds of glass, chromatic aberration may be reduced, but it can never be completely cured. In a refracting telescope

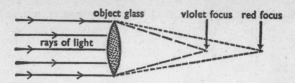

Fig. 9. *Principle of chromatic aberration (not to scale)*

or *refractor, a main lens or *object-glass made in this way is known as an *achromatic object-glass*.

Chromosphere. The part of the Sun's atmosphere lying above the bright surface or *photosphere, and below the outer *corona. It is visible with the naked eye only during a total solar *eclipse, though by means of special instruments it may be observed telescopically at any time. It is red in colour, and is composed chiefly of hydrogen gas.

Chronometer. A very accurate form of timekeeper. Every ship carries a chronometer, without which accurate navigation would be almost impossible.

Circumpolar Star. A star which never sets, but merely circles the pole above the horizon. From Britain, the Great Bear is circumpolar, because it never dips out of view, whereas the brilliant orange star Arcturus is not; the diagram (Fig. 10)

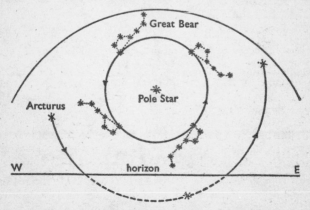

Fig. 10. *Circumpolar and non-circumpolar stars*

will make the situation clear. From more southerly latitudes, however, the Great Bear is not circumpolar, and from countries such as New Zealand it is never visible at all. To New Zealanders the Bear never rises above the horizon; on the other hand, the famous groups near the south celestial pole are circumpolar.

Clark, Alvan (1808–1887). American optician who manufactured many of the largest object-glasses, notably the *Yerkes 40-inch.

Cluster-Cepheids. Obsolete name for *RR Lyræ variables.

Clusters, Star. As the name suggests, a cluster is a group of stars whose members are genuinely associated. They are of two main types: *open* and *globular*.

The open clusters are widely spread, and are made up of a few hundreds of stars, usually together with gas and dust. The most famous of them is the *Pleiades, in the constellation of Taurus (the Bull), which is a conspicuous object with the naked eye; keen-sighted people can see seven or eight separate stars, and binoculars reveal many more. Also in Taurus lies the cluster of the Hyades, which is more scattered and therefore less impressive. The bright orange-red star Aldebaran seems to lie in the Hyades, but in fact it is not a true member of the cluster, and simply happens to be situated in almost the same direction as seen from Earth. The third famous naked-eye cluster is Præsepe, or the 'Beehive', in Cancer (the Crab).

Globular clusters are regular in shape, and contain many thousands of stars; they are very remote, and form a sort of 'outer framework' to the *Galaxy. Only three of them are visible to the naked eye. One is the globular cluster in Hercules, Messier 13; the other two, Omega Centauri and 47 Tucanæ, are too far south in the sky to be seen from Britain or North America. External galaxies, such as the Andromeda Spiral, also contain globular clusters.

Coal Sack. The famous dark *nebula in Crux Australis (the Southern Cross). It is too far south to be visible from Europe.

Cœlostat. A form of optical instrument which makes use of two mirrors, one fixed and one movable (Fig. 11). The movable mirror is mounted parallel to the Earth's axis, and is rotated so that the light from the object under observation is reflected in a fixed direction on to the second mirror. The result of this arrangement is that the eyepiece does not have to move at all.

Strictly speaking, the term 'cœlostat' applies to this moving-mirror device, but it has many applications. It is used, for example, in *tower telescopes, designed for studying the Sun, in which the cœlostat is placed at the top of the tower and the image of the Sun is formed at the base, always in the same position.

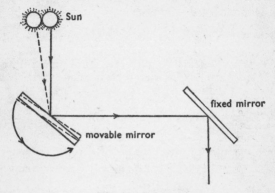

Fig. 11. *Cœlostat*

Collapsar. The end product of a very massive star which has suffered gravitational collapse, so that its escape velocity has become greater than the velocity of light. A collapsar cannot be seen, and is surrounded by a 'forbidden area' known as a *black hole. It must however be stressed that the existence of collapsars (and black holes) has yet to be conclusively proved.

Colour Index. A measure of a star's colour, and hence its surface temperature. The ordinary or *visual magnitude* of a star is a measure of its apparent brightness as seen with the eye; the *photographic magnitude* is obtained by measuring the apparent size of the star's image on a photographic plate. The two magnitudes will not generally be the same, because different colours affect the sensitive plate to different extents; red stars will seem fainter photographically than they appear to the eye. The difference between visual and photographic magnitude is the colour index. The scale is adjusted so that for a white star such as Sirius, of spectral type Ao, colour index is zero.

Consider, for instance, Betelgeux in Orion. The visual magnitude is 0·9 (though admittedly somewhat variable); the photographic magnitude is only 2·6. The difference between 2·6 and 0·9=1·7, which is therefore the colour index of Betelgeux. Since the visual magnitude is the brighter, Betelgeux is shown to be reddish, and the colour index is said to be positive.

Colures. Great circles on the *celestial sphere. The *equinoctial colure* is the great circle of *right ascension 0 hours and 12 hours; it passes through the celestial poles, the *First Point of Aries, and 180 degrees of celestial longitude. The *solstitial colure* is the great circle of right ascension 6 hours and 18 hours, passing through the celestial poles, the poles of the *ecliptic, and the solstitial points (celestial longitude 90 degrees and 270 degrees, or right ascension 6 hours and 18 hours, and *declination $23\frac{1}{2}$ degrees north and south).

Coma. There are two astronomical meanings here:

(1) The hazy-looking patch surrounding the nucleus of a *comet.

(2) The blurred haze surrounding the images of stars on a photographic plate, due to optical defects in the instruments.

Comes (Companion). The fainter member of a *binary.

Comet. A member of the Solar System, moving round the Sun in an orbit which is generally very eccentric.

A comet is not a hard, solid body. It is made up of relatively small particles together with gas and dust, and compared with a planet its mass is very slight. A 'great comet' is an impressive sight, made up of a head with a sharp central nucleus and surrounding *coma, together with a long tail which may stretch for a considerable distance across the sky (Fig. 12). The tail is composed of extremely small particles, and always points more or less away from the Sun, so that when a comet is moving outward it travels tail-first.

However, not all comets have tails, and the average telescopic comet looks like nothing more than a patch of luminous haze in the sky. It has also been found that a comet develops a tail only when it approaches the Sun, and loses its tail again as it recedes. Of course, comets shine because of the radiation they receive from the Sun; they reflect the sunlight, though they may also be caused to emit feeble luminosity on their own account.

Many comets move round the Sun in short periods, of from less than four years up to a few tens of years, but with one exception all these are too faint to be seen with the naked eye.

Fig. 12. *Structure of a large comet*

Most of them have their *aphelia further away from the Sun than the orbit of Jupiter, though when at their closest they may come well within the orbit of the Earth. The average short-period comet cannot be followed continuously, since it is too faint to be seen except when reasonably close to the Sun and to the Earth; but their orbits are known, and one such comet, Encke's, has now been recovered at 50 separate returns.

*Halley's Comet is the only bright comet with a period of less than a century; it takes 76 years to complete one journey round the Sun, and will next return in 1986. All other brilliant comets have periods of hundreds or thousands of years, so that they cannot be predicted.

Quite a number were seen at various times during the Victorian era, but the present century has been depressingly barren of them, though there have been a few bright comets (notably the Daylight Comet of 1910, not to be confused with Halley's, Comet Arend-Roland of April 1957 and Comet Bennett of 1970). Kohoutek's Comet of the winter of 1972–3 was a great disappointment. It was expected to become brilliant, but signally failed to do so, though it was clearly visible with the naked eye. We will hardly be able to study its next appearance, since it will not return to the neighbourhood of the Sun for about 75,000 years!

It is thought that comets are comparatively short-lived, and several former short-period comets seen during the 19th century have now disintegrated. According to the American astronomer F. L. Whipple, the particles making up a comet are mainly of an icy nature.

Comets were once thought to be unlucky, and even dangerous, but it is now known that they are completely harmless; on more than one occasion the Earth has passed through a comet's tail without suffering the slightest damage. There is a close association between comets and *meteors.

Comets are usually named after their discoverers, though

31

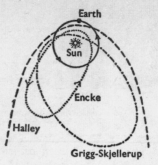

Fig. 13. *Comet orbits*

occasionally after the astronomers who have first worked out
their orbits (Fig. 13). The following is a list of the most interesting
comets with periods of less than 100 years:

Comet	Period (years)	Comet	Period (years)
Encke	3·3	Comas Solá	8·6
Grigg-Skiellerup	4·9	Väisälä	10·5
Pons-Winnecke	6·2	Tuttle	13·6
Kopff	6·3	Crommelin	27·9
Giacobini-Zinner	6·4	Stephan-Oterma	39·0
D'Arrest	6·7	Westphal	61·7
Borrelly	7·0	Brorsen-Metcalf	69·1
Faye	7·4	Olbers	69·6
Whipple	7·5	Pons-Brooks	70·9
Schaumasse	8·2	Halley	76·0

Conjunction. Here again there are several meanings of the term
to be considered.

A planet is said to be in conjunction with a star when it passes
close by; of course, the planet is much closer to us than the star,
and the effect is purely one of line-of-sight. Planets may also be
in conjunction with each other, or with the Moon.

The two planets *Mercury and *Venus, which are closer to
the Sun than we are, are said to be at *inferior conjunction* when
they are more or less between the Sun and the Earth, and at
superior conjunction when they are on the far side of the Sun
(Fig. 14); the lining-up is not usually exact, because both Mercury
and Venus have orbits which are appreciably tilted or inclined
to that of the Earth. An exact lining-up at inferior conjunction
results in a *transit.

The remaining planets lie outside the orbit of the Earth, and
are said to be at superior conjunction when on the far side of the

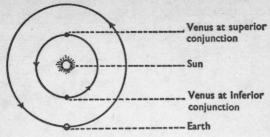

Fig. 14. *Conjunctions of Venus*

Sun, so that they are above the horizon only during broad daylight and are to all intents and purposes unobservable.

Constellation. A group of stars named after a living or mythological character, or an inanimate object; thus there are constellations known as Taurus (the Bull), Orion (the mythological hunter) and Triangulum (the Triangle). The names are generally used in their Latin forms, so that, for instance, the Great Bear becomes Ursa Major.

The original constellations were formed by the ancient stargazers, and are still to be found on modern maps, though in somewhat altered form. Ptolemy, last of the great astronomers of Classical times, listed 48 constellations, among which were the familiar Orion, Ursa Major and others. All Ptolemy's groups are still in use, but the list has been increased since, because Ptolemy's maps did not cover the whole sky; in particular, he knew nothing about the stars near the south celestial pole, which are never visible from the latitude of his home at Alexandria.

It is important to note that a constellation is not made up of stars which are genuinely associated with each other. In Ursa Major, for example, the end star in the Bear's tail, Alkaid, is 210 *light-years away from us, while Mizar, the second star, is only 88 light-years distant; thus Alkaid is considerably further from Mizar than we are. Were we observing from a different position in the *Galaxy, our whole view would be different.

It cannot be said that many of the constellations have outlines which give much impression of the objects they are supposed to represent. Ursa Major is nothing like a bear, and its nicknames of the Plough and (in America) the Big Dipper are more appropriate. Triangulum (the Triangle) is one of the few exceptions.

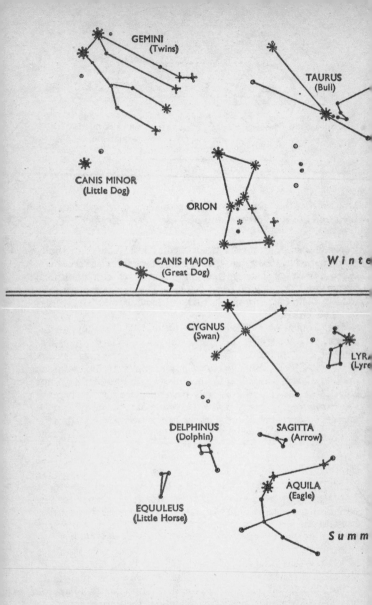

GEMINI
(Twins)

TAURUS
(Bull)

CANIS MINOR
(Little Dog)

ORION

CANIS MAJOR
(Great Dog)

Winter

CYGNUS
(Swan)

LYRA
(Lyre)

DELPHINUS
(Dolphin)

SAGITTA
(Arrow)

AQUILA
(Eagle)

EQUULEUS
(Little Horse)

Summer

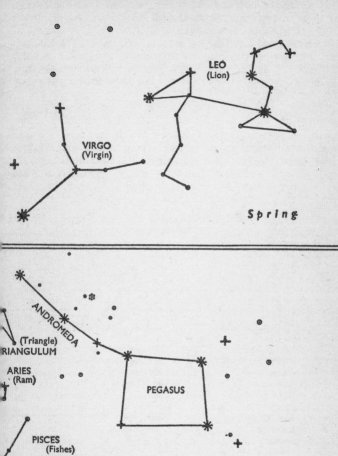

Fig. 15. *Principal constellations in the four seasons*

Because the stars are so remote, their individual or *proper motions are very slight, even though the stars are really moving through space at all sorts of speeds in all sorts of directions. Therefore, the constellation patterns seem to stay unchanged for long periods of time, and have not altered markedly since the dawn of civilization (Fig. 15). Eventually, however, the slow individual movements of the stars will cause the constellations to change in outline.

The following constellations are now recognized by the International Astronomical Union. The huge, unwieldy constellation of Argo Navis (the Ship Argo) has been officially divided up into Carina (the Keel), Vela (the Sails) and Puppis (the Poop). Large and important constellations are given in capital letters, while constellations partly or wholly invisible in Britain and North America are marked with the sign §.

ANDROMEDA	Andromeda
§Antila	The Air-pump
§A	The Bee
Aquarius	The Water-bearer
AQUILA	The Eagle
§Ara	The Altar
§ARGO NAVIS	The Ship Argo
Aries	The Ram
AURIGA	The Charioteer
BOÖTES	The Herdsman
§Cælum	The Sculptor's Tools
Camelopardus	The Giraffe (Camelopard)
Cancer	The Crab
Canes Venatici	The Hunting Dogs
CANIS MAJOR	The Great Dog
Canis Minor	The Little Dog
Capricornus	The Sea-Goat
§CARINA	The Keel
CASSIOPEIA	Cassiopeia
§CENTAURUS	The Centaur
Cepheus	Cepheus
Cetus	The Whale
§Chamæleon	The Chameleon
§Circinus	The Compasses
§Columba	The Dove
Coma Berenices	Berenice's Hair
§Corona Australis	The Southern Crown
Corona Borealis	The Northern Crown
Corvus	The Crow
Crater	The Cup
§CRUX AUSTRALIS	The Southern Cross
CYGNUS	The Swan
Delphinus	The Dolphin
§Dorado	The Swordfish

Draco	The Dragon
Equuleus	The Little Horse
§Fornax	The Furnace
GEMINI	The Twins
§GRUS	The Crane
Hercules	Hercules
§Horologium	The Clock
Hydra	The Sea-Serpent
§Hydrus	The Watersnake
§Indus	The Indian
Lacerta	The Lizard
LEO	The Lion
Leo Minor	The Little Lion
Lepus	The Hare
Libra	The Balance (Scales)
§LUPUS	The Wolf
Lynx	The Lynx
LYRA	The Harp (Lyre)
§Mensa	The Table
§Microscopium	The Microscope
Monoceros	The Unicorn
§Musca Australis	The Southern Fly
§Norma	The Rule
§Octans	The Octant
Ophiuchus	The Serpent-Bearer
ORION	Orion
§Pavo	The Peacock
PEGASUS	The Flying Horse
PERSEUS	Perseus
§Phœnix	The Phœnix
§Pictor	The Painter
Pisces	The Fishes
Piscis Austrinus	The Southern Fish
§PUPPIS	The Poop
§Reticulum	The Net
Sagitta	The Arrow
SAGITTARIUS	The Archer
SCORPIO (or Scorpius)	The Scorpion
Sculptor	The Sculptor
Scutum	The Shield
Serpens	The Serpent
Sextans	The Sextant
TAURUS	The Bull
§Telescopium	The Telescope
Triangulum	The Triangle
§Triangulum Australe	The Southern Triangle
§Tucana	The Toucan
URSA MAJOR	The Great Bear
Ursa Minor	The Little Bear
§VELA	The Sails
VIRGO	The Virgin
§Volans	The Flying-Fish
Vulpecula	The Fox

Copernican System. In 1543, the Polish scientist Nicolaus Copernicus (1473–1543) published a book in which he claimed that instead of the Sun and planets moving round the Earth, the Sun lies at the centre of the *Solar System, with the Earth and other planets revolving round it. Though Copernicus's ideas were inaccurate in many ways, he was of course correct in placing the Sun in the central position, and his scheme has become known as the Copernican system. It was hotly challenged at the time, largely on religious grounds, and was not universally accepted until more than a century after Copernicus's death.

Corona. The outermost part of the Sun's atmosphere; it is made up of very tenuous gas at a surprisingly high temperature, and is of great extent. It is visible to the naked eye only during a total solar *eclipse. Using a special instrument known as a *coronagraph* (invented by the French astronomer B. Lyot) its inner parts may be studied without the help of an eclipse, but such observations are far from easy.

Coronagraph. See under the heading *Corona*.

Cosmic Rays. High-speed particles reaching the Earth from outer space. Their origin is still not definitely known; though it has been found that the Sun is a minor source, most of the cosmic radiation comes from outside the Solar System.
  The heaviest cosmic-ray particles are broken up when they enter the top part of the Earth's *atmosphere, so that at ground level we are shielded from them. Modern cosmic-ray studies are carried out largely by means of instruments carried in *artificial satellites or interplanetary probes.

Cosmogony. The branch of science dealing with the origin and development of the *universe, or of any particular part of it.

Cosmology. The study of the *universe as a whole; its nature, and the relations between its various parts.

Counterglow. The English name for the glow known more commonly by its German title, the *Gegenschein.

Crab Nebula. A gas-cloud in the constellation of Taurus (the Bull), near the third-magnitude star Zeta Tauri. It is much too faint to be seen with the naked eye, but a small telescope will show it as a dim, misty patch; photographs taken with large instruments reveal a very complex structure.

The Crab Nebula is 6000 light-years away. It is one of the strongest radio sources in the sky; it is also a source of X-rays, and it contains a *pulsar, which has been aptly described as the 'power-house' of the Nebula. In our experience it is unique, and we know it to be the remains of the brilliant *supernova which was observed by Chinese astronomers in the year 1054, and which became bright enough to be visible with the naked eye in broad daylight. All that is left is the pulsar, together with the still-expanding cloud of gas. So much information has been drawn from it that it has even been said that there are two types of astronomy: the astronomy of the Crab Nebula, and the astronomy of everything else!

Crimean Astrophysical Observatory. Leading Russian observatory; the largest telescope is the 102-inch reflector. The present Director is A. Severny.

Crommelin, Andrew Claude de la Cherois (1865–1939). English astronomer, noted for his work on the orbits of comets.

Crux Australis. The Southern Cross, most famous of all the southern constellations. It is too far south to be seen from Europe. Crux is actually the smallest constellation in the entire sky!

Culmination. The maximum altitude of a celestial body above the horizon. The Sun, of course, culminates at noon.

D

Darwin, Sir George (1845–1912). Son of Charles Darwin. He is best remembered for his tidal theory of the origin of the Moon.

Dawes's Limit. The practical limit for the *resolving power of a telescope; it is $\frac{4\cdot56}{d}$, where d is the aperture of the telescope in inches.

Day. The period taken for the Earth to spin once on its axis. However, there are various kinds of 'days'.

A *sidereal day* is the time interval between successive meridian passages, or *culminations, of the same star (23h 56m 4s.091). In other words, the Earth's rotation period is measured with reference to the stars, which are so far away that for all practical purposes we may regard them as infinitely remote.

A *solar day* is the time interval between two successive noons. It is slightly longer than the sidereal day, since the Sun is moving against the starry background at about one degree per day in an easterly direction. The situation is complicated by the fact that the Earth's orbit is somewhat elliptical; when at its nearest to the Sun, the Earth moves at its fastest, according to the principles of *Kepler's Laws, and so the Sun seems to move at its fastest against the stars. (The stars cannot be seen with the naked eye during daylight only because they are overpowered by the brilliance of the Sun and the sky.) For convenience, astronomers normally measure by a *mean sun*, which is an imaginary body moving round the celestial equator at a constant speed which is equal to the average rate of motion of the true Sun around the *ecliptic. The *mean solar day* is 24h 3m 56s.555 long.

Astronomically, the 24-hour clock is used, midnight being 0 hours. (In a *civil day* there are two 12-hour periods, a.m. and p.m.) For scientific purposes, daylight-saving adjustments, such as British Summer Time, are ignored; everything is recorded in *Greenwich Mean Time.

Declination. The angular distance of a celestial body north or south of the celestial *equator. Therefore, the equator itself has a declination of 0°, while the north celestial pole is at 90°N. or +90°; Polaris, with a declination of +89°3′, is less than a degree away. Declination in the sky corresponds to latitude on the Earth.

Deferent. In Ptolemy's system of the universe, all the bodies in the sky moved round the Earth in circular orbits; the circle was said to be the 'perfect' form, and nothing short of perfection could be allowed in the heavens! Unfortunately it was obvious that the movements of the planets could not be explained by supposing that they moved in circular paths at constant velocities. Ptolemy, who was an excellent observer and mathematician, knew this quite well, so he overcame the difficulty by supposing that a planet moved round the Earth in a small circle or *epicycle*, the centre of which – the deferent – itself moved round the Earth in a perfect circle (Fig. 16). As more and more irregularities came to light, more and more epicycles had to be introduced, until the whole theory became hopelessly clumsy and artificial. All the same, it lasted for many centuries, and it was only in the years following 1543 that it was replaced by the *Copernican System.

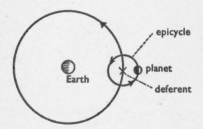

Fig. 16. *Ptolemaic epicycle and deferent*

Degree of Arc. A unit for measuring angles. A full circle contains 360 degrees (Fig. 17); therefore, a right angle contains one-quarter of 360 degrees (=90 degrees). Each degree is subdivided into 60 minutes of arc, and each minute is again subdivided into 60 seconds of arc. The symbols are ° (degree), ′ (minute) and ″ (second).

Deimos. The outer satellite of Mars; it is an irregularly-shaped body with a cratered surface, as was shown by photographs from Mariner 9. It has a diameter of less than 15 miles. See *Satellites.

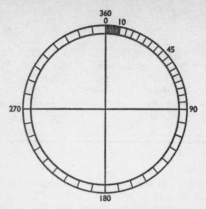

Fig. 17. *Degrees in a circle*

Delta Cephei. The prototype *Cepheid. It has a period of 5·3 days.

Density. The amount of matter in a unit volume of a substance. For most purposes, water is taken as 1; thus the mean density of the Earth is 5·52, i.e. a given volume of the Earth weighs 5·52 times as much as an equal volume of water would do. (It must be remembered that this relates to the Earth as a whole. Near the outer crust, the density is well below 5; near the centre of the Earth's heavy core, the value may rise to as much as 9 or 10.) The densities of the other planets are given on page 25. Taking water as unity, the Sun's mean density is only 1·4. On the other hand, the densities of some of the curious *White Dwarf stars are extremely high – 30,000 in the case of the faint companion of Sirius – and the *neutron stars, or *pulsars, are much denser still.

Descending Node. This is described under the heading *Nodes.

De Sitter, Willem (1872–1935). Dutch astronomer, and a leading cosmologist. He was Director of the Leiden Observatory from 1919 to 1935.

Dewcap. An open tube fitted to the end of a refracting telescope or *refractor, beyond the *object-glass (Fig. 18). It prevents the

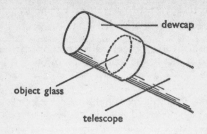

Fig. 18. *Dewcap*

object-glass from becoming moistened by condensation – at least, it should do so!

Dichotomy. The exact half-*phase of Mercury, Venus or the Moon.

Diffraction Grating. A device used for splitting up light; it may be regarded as an alternative to the *prism of a *spectroscope, and is now widely used. The grating consists of a series of close parallel lines ruled on a polished metallic surface. The lines must be very close indeed – several thousands of them to each inch – so that diffraction gratings are not easy to make.

Diffraction Rings. Concentric rings surrounding the image of a star as seen in a telescope. They cannot be eliminated, since they are due to the wave-motion of light and the construction of the telescope; however, they are more of a nuisance in small telescopes than in large ones.

Dione. The fourth *satellite of Saturn.

Dipper. American unofficial name for *Ursa Major (the Great Bear).

Direct Motion. Bodies which move round the Sun in the same sense as the Earth are said to have direct motion; those which move the opposite way have *retrograde motion. The terms may also be used with regard to satellites. No planet or minor planet with retrograde motion is known, but there are various retrograde comets and satellites.

The terms are also used with regard to the apparent move-

43

ments of the planets in the sky. When shifting eastward among the stars, a planet is said to have direct motion; when moving westward, it is retrograding.

Diurnal Motion. The apparent daily rotation of the sky from east to west. It is due to the real rotation of the Earth from west to east.

Doppler Effect. The apparent change in wavelength of light (or sound) caused by the motion of the source or of the observer.

The classical example is that of a whistling train by-passing an observer. When the train is approaching, more sound-waves per second enter the observer's ear than would be the case if the train were standing still; the wavelength is shortened, and the whistle is high-pitched. When the train starts to recede, fewer sound-waves per second enter the ear; the wavelength is lengthened, and the note of the whistle drops. It is much the same with light, but here the effect is to make the object 'too blue' when approaching and 'too red' when receding (Fig. 19).

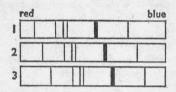

Fig. 19. *The Doppler Shift*
1. *Receding*   2. *Stationary*   3. *Approaching*

The actual changes in colour are extremely slight, but the effects of the motion show up in the spectra of the light-sources. With stars, for instance, all the spectral lines are shifted *en bloc*, to the red or long-wave end if the star is receding, to the blue or short-wave end in the case of a velocity of approach. All the external *galaxies, apart from those of our own *Local Group, show *red shifts. This is taken to mean that all these galaxies are racing away from us, and that the entire universe is expanding.

The Doppler Effect is named in honour of the Austrian physicist who first drew attention to it in the year 1842. It is of fundamental importance in astronomical studies.

Double Star. A star made up of two components relatively close together. Some doubles are *optical*; that is to say, the components are not truly associated, and simply happen to lie in the same direction as seen from the Earth. Most, however, are *binary systems, in which the association is real.

Draper, Henry (1837–1882). American pioneer of astronomical photography.

Dreyer, J. L. E. (1852–1926). Danish astronomer. He spent many years in Ireland, first with Lord *Rosse and then as Director of the *Armagh Observatory. He compiled the New General Catalogue of clusters and nebulæ, and was also a great astronomical historian.

Driving Clock. A mechanical device used to move a telescope round just fast enough for it to follow a celestial body moving from east to west across the sky (see Diurnal Motion). Electrical drives are now generally used, but are still often termed 'clocks'.

Generally speaking, a drive can be used only when the telescope is fitted with an *equatorial mounting, though it is true that the new Russian 236-inch reflector has an altazimuth mount and a correspondingly more elaborate driving system.

Dunsink Observatory. The main observatory in Eire, five miles from Dublin. It was founded in 1785. The present Director is P. A. Wayman.

Dwarf Novæ. Common term for *U Geminorum variable stars.

Dyson, Sir Frank (1868–1939). The ninth Astronomer Royal (1910–33).

E

Earth. The third planet in order of distance from the Sun. Details of its orbit and dimensions will be found under the heading *Planets.

The mean distance between the Earth and the Sun is 92,957,209 miles, but since the orbit is not a perfect circle the distance ranges between 91,400,000 miles at *perihelion to 94,600,000 miles at *aphelion. The *seasons are due not to this changing distance, but to the fact that the Earth's axis is tilted from the perpendicular by an angle of 23½°.

The Earth is the largest of the four inner planets, though Venus is almost its equal. Its mass is some 6,000,000,000,000,000,000,000 tons, and its mean *density is 5·52 times that of water. There is presumably a heavy core, but we have to admit that our knowledge about the central regions of the globe is extremely slight. The *atmosphere is made up chiefly of nitrogen (77·6%) and oxygen (20·7%); no other planet in the Solar System has nearly so much free oxygen in its atmosphere, so that there is nowhere else where an Earth-type man or animal could breathe.

In shape, the Earth is not a perfect sphere. Its diameter is 7926 miles as measured through the equator, but only 7900 miles as measured through the poles. This is because the axial rotation is fairly rapid (approximately 24 hours), causing the equatorial regions to 'bulge' slightly.

The age of the Earth is about 4,700,000,000 years. This figure is probably fairly accurate, but we are by no means certain how the Earth and other planets were formed – a problem discussed briefly under the heading *Solar System.

Earthshine. When the Moon appears as a crescent, the 'dark' side may often be seen shining dimly. This is because the night side of the Moon is being lit up by light reflected on to it by the Earth.

Eclipses, Lunar. Because the Moon revolves round the Earth (or, to be more accurate, because both bodies move round their common centre of gravity or *barycentre), there must be times when the Moon passes into the shadow cast by the Earth. This causes an eclipse of the Moon, or lunar eclipse (Fig. 20).

During an eclipse, the Moon does not vanish completely;

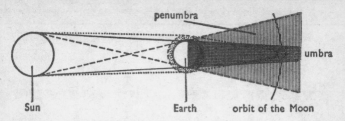

Fig. 20. *Theory of a lunar eclipse*

some sunlight is bent or refracted on to its surface by the Earth's atmosphere, as shown in the diagram, so that the Moon merely turns a dim, often coppery colour until it passes out of the shadow again. Lunar eclipses may be either total, when the whole of the Moon passes into the shadow, or partial, when only a portion of the Moon is covered. Obviously, a lunar eclipse can happen only at full moon; the reason why an eclipse is not seen every month is because the Moon's orbit is tilted relative to that of the Earth, and no eclipse can take place unless the full moon is close to a *node. Totality can never last for more than 1¾ hours.

Since the Sun is a disk, and not a point source of light, there is an area of partial shadow or *penumbra* to either side of the main cone of shadow, or *umbra*, cast by the Earth. If the Moon misses the main shadow, but enters the penumbra, a *penumbral eclipse* results, and the slight dimming may just be noticed with the naked eye. Of course, the Moon must always pass through the penumbra before entering the main cone.

As soon as the sunlight is cut off from the Moon, the lunar surface temperature drops sharply; it may fall by over 200 degrees Fahrenheit in a single hour, showing that the Moon's crust is very poor at retaining heat. On the other hand, there are some areas on the Moon – notably the region of the conspicuous, 54-mile crater Tycho – which cool down more slowly, and which must therefore have a different sort of surface coating, or roughness. These regions are known, rather misleadingly, as 'hot spots'.

Eclipses of the Moon are not of great importance astronomically, but temperature-studies have given us useful information about the nature of the crust. As a spectacle, an eclipse is well worth watching, since beautiful coloured glows on the Moon may sometimes be seen; these are, of course, due to the light which has passed through the Earth's air, and they have nothing to do with the Moon itself.

47

If a lunar eclipse occurs, it is visible over an entire hemisphere of the Earth. Therefore, any particular place on the Earth will see eclipses of the Moon more frequently than eclipses of the Sun.

The following lunar eclipses will take place during the period 1975–1980:

Eclipses wholly or partly visible from Great Britain.	Eclipses wholly or partly visible from N. America.
1975 Nov. 18 (Total)	1975 May 25 (Total)
1976 May 13 (Small partial)	1975 Nov. 18 (Total)
1977 Apr. 4 (Small partial)	1977 Apr. 4 (Small partial)
1978 Sept. 16 (Total)	1979 Sept. 6 (Total)
1979 Mar. 13 (Large partial)	

Eclipses, Solar. By a lucky chance, the Sun and Moon appear almost the same size in the sky. Therefore, the Moon may sometimes pass in front of the Sun, hiding or eclipsing it (Fig. 21).

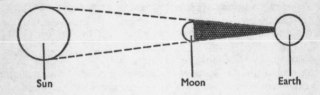

Fig. 21. *Theory of solar eclipse*

Solar eclipses are of three kinds. In a *total eclipse*, the Sun is completely hidden, and the effect is magnificent, since the solar *corona and *prominences flash into view. If the eclipse is *partial*, neither the corona nor prominences may be seen with the naked eye. If exact lining-up occurs when the Moon is at its greatest distance from the Earth, the lunar disk appears slightly smaller than that of the Sun, and cannot fully cover it, so that a ring of sunlight is left showing round the dark, invisible body of the Moon. This is known as an *annular eclipse*.

Obviously, a solar eclipse can happen only at new moon. The 5°9′ tilt of the Moon's orbit with respect to that of the Earth means that eclipses do not happen each month; usually the new moon passes above or below the Sun in the sky. Moreover, a solar eclipse is not visible over a full hemisphere of the Earth, as with an eclipse of the Moon.

Total eclipses are of great importance to astronomers, since there are various investigations which can be carried out only when the Sun is hidden by the Moon. Unfortunately they are not so common as might be wished, and the track of totality can never be more than 169 miles wide; totality itself can never last for more than about 8 minutes, and is generally much shorter even than this. The next total eclipse visible from England will not take place until 1999 August 11, when the track will cross Cornwall.

The following eclipses will occur during the period 1975–1980:

Total – 1976 Oct. 23 (S. Indian Ocean area). 1977 Oct. 12 (Pacific area). 1980 Feb. 16 (East African area).

Annular – 1976 Apr. 29 (Libya, Turkey etc.). 1977 Apr. 18 (South-West African area). 1979 Aug. 22 (Antarctica). 1980 Aug. 10 (South Pacific, Brazil, etc.).

Partial – 1975 May 11 (Arctic area; small partial seen from Britain). 1975 Nov. 3 (Antarctica). 1978 Apr. 7 (Antarctica). 1978 Oct. 2 (Arctic).

Of course, a total or annular eclipse will be seen as a partial to either side of the central track.

Eclipsing Variable (or Eclipsing Binary). A *binary star, made up of two components moving round their common centre of gravity in such a way that one star passes periodically in front of the other as seen from Earth, so producing an eclipse. When this happens, the total light seems to fade (Fig. 22).

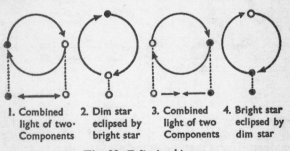

1. Combined light of two· Components 2. Dim star eclipsed by bright star 3. Combined light of two Components 4. Bright star eclipsed by dim star

Fig. 22. *Eclipsing binary*

The best-known member of the class is Algol, in the constellation Perseus. Usually the *magnitude is about equal to that of the Pole Star (2·0); but when the fainter member of the Algol system passes in front of the brighter, the magnitude drops to below 3. Minimum lasts for more than a quarter of an hour before the main eclipse is over; as the dim companion moves aside, the magnitude increases again to its normal value. Many Algol-type stars have been found, though few are visible to the naked eye.

In another famous eclipsing variable, Beta Lyræ (near Vega), the two components are less unequal; they are extremely close together, and each must be drawn out into an egg-like shape, though they cannot, of course, be seen separately. With Beta Lyræ there are two minima in each full period, one minimum being fainter than the other.

Ecliptic. The projection of the Earth's orbit on to the *celestial sphere. It may also be defined as 'the apparent yearly path of the Sun among the stars', passing through the twelve constellations of the *Zodiac. Since the plane of the Earth's orbit is inclined to the equator by an angle of $23\frac{1}{2}$ degrees, the angle between the ecliptic and the celestial *equator must also be $23\frac{1}{2}$ degrees.

Eddington, Sir Arthur Stanley (1882–1944). English astronomer; one of the greatest of all cosmologists, who also undertook pioneer theoretical work into the constitution and evolution of the stars.

Einstein, Albert (1879–1955). German scientist who laid down the principles of relativity theory; probably the greatest mathematician since Newton. He lived in America from 1933 until his death.

Elara. The seventh satellite of Jupiter.

Electromagnetic Spectrum. The full range of what is termed *electromagnetic radiation*: radio waves, visible light, and very short radiations such as X-rays. As is made clear by the diagram (Fig. 23), visible light takes up only a very small part of the whole electromagnetic spectrum.

Electron. A fundamental particle carrying a negative charge; it makes up part of an *atom. Each electron is almost inconceivably small, and the number needed to make a weight of one ounce has been given as 311×10^{26} – in other words, 311 followed by 26 zeros. However, it is misleading to picture an electron as being a solid lump.

Electron Density. The number of free electrons in a unit volume of space. (A free electron is an electron which is not attached to any particular atom, but is moving around on its own.)

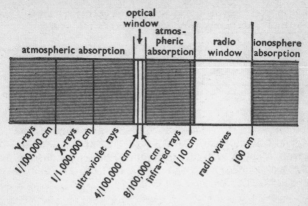

Fig. 23. *The electromagnetic spectrum*

Element. A substance which cannot be chemically split up into simpler substances. It may be said that the elements are the fundamental 'building bricks' of the universe. Familiar elements include hydrogen, helium, oxygen, iron, gold, silver, mercury and tin. All other substances are made up of these elements – for instance, water is made up of hydrogen and oxygen.

Elongation. The apparent angular distance of a planet from the Sun, or of a satellite from its parent planet. When a planet is at *opposition, and so is directly opposite to the Sun in the sky, its elongation is 180 degrees. This can never apply to Mercury or Venus, which are closer to the Sun than we are, and are always to be found somewhere near the Sun in the sky; the maximum elongation is 47 degrees for Venus, only 28 degrees for Mercury.

Enceladus. The second *satellite of Saturn.

Encke, Johann Franz (1791–1865). German astronomer; Director of the Berlin Observatory. He calculated the distance of the Sun from observations of the *transits of Venus of 1761 and 1769, and discovered the periodicity of the *comet which now bears his name.

Encke's Comet. A short-period comet; it takes only 3·3 years to move round the Sun, and has now been seen at over 50 returns.

51

Encke's Division. A 'ripple' in Saturn's outer ring; probably not a true division comparable with *Cassini's Division.

Ephemeris. A table showing the predicted positions of a moving celestial body, such as a planet or a comet.

Epicycle. This is described under the heading *Deferent.

Epoch. A date chosen for reference purposes in quoting astronomical data. For instance, some star catalogues give the apparent positions of the stars (in right ascension and declination) for 'epoch 1950'. By the year 2000 the positions will have changed slightly, because of the *precession of the equinoxes, so that adjustments will have to be made.

Equation of Time. As explained under the heading *Day, the Sun does not move among the stars at a constant speed, because the Earth's orbit is not a perfect circle. Astronomers therefore make use of a *mean sun, an imaginary body which travels round the celestial equator at a speed equal to the average speed of the real Sun. The interval by which the real Sun is ahead of or behind the mean sun is termed the equation of time. It can never be greater than 17 minutes; four times every year it becomes zero, so that the right ascension of the mean Sun is then the same as the right ascension of the real Sun.

Equator, Celestial. The projection of the Earth's equator on to the *celestial sphere. It divides the sky into two equal parts, a northern hemisphere and a southern hemisphere. Delta Orionis, the northernmost of the three bright stars forming the Belt of Orion, lies very close to the celestial equator, so that its declination is practically 0 degrees.

Equatorial Mount. If a telescope is mounted upon an axis which is parallel to the axis of the Earth, it need be moved only in right ascension (east to west) in order to keep a celestial body in the field of view; the up-or-down movement (declination) will look after itself. The polar axis, upon which the telescope is mounted, points to the celestial pole, which so far as the northern hemisphere is concerned is marked approximately by Polaris. There are various types of equatorial mounts, but all depend upon the same principle.

An equatorial mount is much more convenient than a simple *altazimuth, and is used for all large astronomical telescopes.

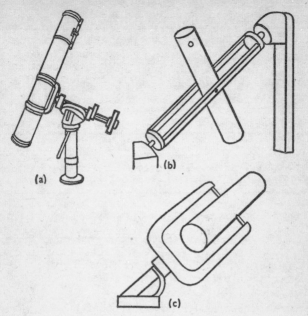

Fig. 24. *Equatorial mounts*
(a) *German* (b) *English* (c) *open fork*

If a *driving clock is to be fitted, an equatorial mounting is essential.

Various forms of equatorials are shown in the diagram (Fig. 24). With the *German* type, a second shaft extends at right angles from the polar axis; one side takes the telescope, while the other side carries a counterweight. In the *English* type, the telescope is pivoted inside a large yoke, inclined at the correct angle and supported by a pier. Better, perhaps, is the *fork*, which is not unlike the English, but has no pier; the upper part of the yoke is missing, and the telescope tube is pivoted between the prongs of the fork. Finally there is the *Foucault*, in which the polar axis broadens out into a large disk; two stout arms are fixed to the face of the disk, forming the arms between which the telescope is mounted.

Equinox. Twice a year the Sun crosses the celestial equator, once when moving from south to north (about March 21) and once when moving from north to south (about September 22). These points are the two equinoxes; that in March is called the *spring* or *vernal equinox* (First Point of Aries), while that in September is known as the *autumnal equinox* (*First Point of Libra).

Another way of putting it is to say that the equinoxes are the two points at which the ecliptic cuts the celestial equator (Fig. 25).

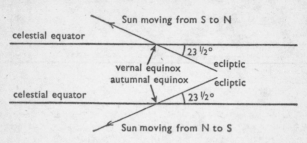

Fig. 25. *Ecliptic and celestial equator*

Eratosthenes (BC 276–196). Born at Cyrene. Greek astronomer. He measured the size of the Earth with amazing accuracy.

Eros. A small *minor planet, of irregular shape; its longest diameter is less than 20 miles. It has a period of 1·76 years. Though its mean distance from the Sun is 135,600,000 miles, it has an eccentric orbit which sometimes brings it within 15,000,000 miles of the Earth (Fig. 26), and during the close approach of 1931 it was closely studied, because measures of its position allowed its orbit to be worked out very accurately – which, in turn, provided a key to the length of the *astronomical unit or Earth-Sun distance. Better ways of measuring the astronomical unit are now used, so that at its last close approach, that of 1974–5, it received rather less attention in this respect, though its movements were closely studied.

Eros rotates in a period of 5h 16m, and this, of course, causes changes in its brightness; it is most brilliant when 'broadside-on' to us. Telescopically it looks like a star, though in 1931 van den Bos and his colleagues in South Africa, using the 27-inch refractor at Johannesburg, were able to see the elongated shape. It may well be pitted with craters in the same manner as the two dwarf satellites of Mars, though of course we have no proof.

Eros was discovered in 1898, photographically, by Witt at Berlin; it was photographed on the same night by Charlois, in France, who however was less quick at comparing his plates and

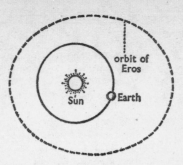

Fig. 26. *Orbit of Eros*

so missed the honour of being co-discoverer. Eros is No. 433 in the list of minor planets. Since it was the first known minor planet to come within the orbit of Mars, it was also the first to be given a masculine name.

Escape Velocity. The minimum velocity at which an object must move in order to escape from the surface of a planet, or other body, without being given any extra propulsion. It is 7 miles per second in the case of the Earth; therefore, a rocket sent upward at 7 miles per second (about 25,000 m.p.h.) will escape, while a rocket sent up at a lesser velocity will fall back. The Moon, which is much less massive than the Earth, has an escape velocity of only $1\frac{1}{2}$ miles per second; for the giant planet Jupiter, on the other hand, the value is as much as 37 miles per second.

Euler, L. (1707–1783). Swiss mathematician, who made outstanding contributions to dynamical astronomy.

Europa. The second Galilean satellite of Jupiter; it is slightly smaller than our Moon. See *Satellites.

Evection. An inequality in the Moon's motion. It is due to the pull of the Sun which affects the orbit of the Moon and makes it alternately a little more eccentric and a little less eccentric. Evection is predictable, and has to be taken into account when working out the times of, say, *eclipses; the effect may amount to as much as 3 hours.

Exosphere. The outermost part of the Earth's *atmosphere. It is very rarefied, and has no definite boundary, since it simply 'thins out' into space.

Extinction. When a star or planet is low in the sky, its light is coming to the observer through a relatively thick layer of atmosphere, and so the brightness is reduced. This effect is known as extinction. It amounts to 3 magnitudes for a star only 1 degree above the horizon, but to only one magnitude for a star at an altitude of 10 degrees. Above 45 degrees altitude, extinction is so slight that it may be neglected for most practical purposes.

Eyepiece (or *Ocular*). The lens, or combination of lenses, placed at the eye-end of a telescope. Its role is to magnify the image formed by the object-glass of a refractor, or the mirror of a reflector; in fact, all the magnification is done by the eyepiece. An astronomical telescope will have several eyepieces, so that different magnifications may be used as desired.

There are various types. With a *positive* eyepiece, the image-plane is outside the eyepiece – between it and the object-glass or mirror – so that it can be used with a *micrometer; examples are the forms known as the Ramsden, Orthoscopic and Mono-centric. With a *negative* eyepiece, such as the Huyghenian or Tolles, the image-plane lies inside the eyepiece. A *Barlow* is a concave lens of about 3 inches negative focal length, mounted in a short tube which can be placed between the object-glass (or mirror) and the eyepiece, inside the draw-tube of the eyepiece. It increases the effective focal length of the object-glass or mirror, and so gives extra magnification.

F

Faculæ. Bright, temporary patches on the surface of the sun. They are often, though not always, associated with sunspots, and are easily visible with a small telescope. Faculæ frequently appear in a position near which a spot-group is about to appear, and may persist for some time in the region of a group which has disappeared.

Filar Micrometer. A device used for measuring very small distances as seen in the eyepiece of a telescope (Fig. 27). In its simplest form, it consists of two thin wires (often spider-threads), one of which is fixed, while the other may be moved with the aid of a fine screw. Filar micrometers are widely used for measuring the distances between the components of visual *binary stars.

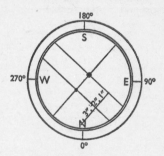

Fig. 27. *Micrometer*

Finder. A small telescope attached to a larger one (Fig. 28). The finder has a low magnification, but it also has a wide field of view, which makes objects easy to locate; it is much more difficult to get an object into the relatively small field of the larger telescope. The procedure is to bring the object to the centre of the finder field; when this has been done, the object will be visible in the larger telescope, always provided that the finder is correctly lined up!

Fireball. An exceptionally bright *meteor.

Fig. 28. *Finder telescope*

First Point of Aries. The point at which the *ecliptic cuts the celestial *equator, with the Sun moving from south to north. The Sun reaches this point about March 21 each year, at the spring or vernal *equinox. In ancient times, the point used to lie in the constellation of Aries (the Ram); the effects of *precession have now shifted it into the neighbouring constellation of Pisces (the Fishes), but the old name is still used. For obvious reasons, the right ascension and declination of the First Point of Aries are both zero.

First Point of Libra. The point at which the *ecliptic cuts the celestial *equator, with the Sun moving from north to south.

Flammarion, Camille (1842–1925). French astronomer, best remembered for his research in connection with Mars and for his popularization of astronomy.

Flamsteed, John (1646–1719). English clergyman, who became the first Astronomer Royal (from 1675 until his death). His main contribution was in drawing up a star catalogue, much the best of its time.

Flares, Solar. Brilliant outbreaks in the outer parts of the Sun's atmosphere, usually associated with active sunspot groups. They are made up of hydrogen, and send out electrified particles which may later reach the Earth, causing *magnetic storms and displays of *auroræ as well as interfering with radio communications. Flares are observed by means of instruments based upon the principle of the *spectroscope. They are not generally visible in ordinary telescopes, but a few exceptional flares have been seen in such a way; in 1965, for instance, a flare was observed by the British amateur astronomer P. Ringsdore, using a 4-inch refractor without any spectroscopic equipment attached.

Flare Stars. Some faint red dwarf stars which may brighten up by several magnitudes over a period of a few minutes, fading back to their usual brightness within an hour or so. It is thought that this behaviour must be due to intense flare activity in the star's atmosphere. The brightest flare star is *UV Ceti.

Flash Spectrum. Just before the Moon completely covers the Sun at a total solar *eclipse, the Sun's atmosphere (the *chromosphere) is seen shining by itself, without the usual brilliant background of the solar surface. The dark lines in its spectrum then become bright, producing the so-called flash spectrum. The same effect occurs just after the end of totality.

Flocculi. Patches on the Sun's surface, observed by instruments based on the principle of the *spectroscope. Bright flocculi are usually made up of calcium; dark flocculi are composed of hydrogen.

Flying Saucers. Unidentified flying objects reported at intervals since 1947. Most of them are due to clouds, ice crystals and weather balloons. They are certainly not space-ships from other planets, as has been claimed in various sensational books!

Fraunhofer, Joseph von (1787–1826). German optician, who mapped the dark lines in the solar spectrum. His object-glasses were the best of their time.

Fraunhofer Lines. The dark absorption lines in the spectrum of the *Sun (Fig. 29).

Fig. 29. *Absorption of Fraunhofer lines*

Focal Length. The distance between a lens (or mirror) and the point at which the image of the object being studied is brought to focus. For instance, a 6-inch mirror may bring an object to focus at a distance of 48 inches beyond the mirror; in this case the focal length is 48 inches, and the f-ratio is 48/6=8, usually abbreviated to f/8.

Forbidden Lines. Certain lines in the spectrum of a celestial body which do not usually appear under normal laboratory conditions, but which may be prominent in the spectra of *nebulæ and other objects.

Free Fall. The normal state of motion of an object in space under the influence of the gravitational pull of a central body; thus the Earth is in free fall round the Sun. Similarly, an *artificial satellite is in free fall round the Earth; a man inside it will have no apparent 'weight', and will be experiencing what is known as *zero gravity.

G

Galaxies. Galaxies, formerly known as *extragalactic nebulæ*, are star-systems; each is made up of many thousands of millions of stars, together with gas and dust. About 1000 million of them are within the photographic range of the world's largest telescope, but only three are clearly visible with the naked eye; the *Andromeda Spiral in the northern hemisphere of the sky, and the two *Nubeculæ or Magellanic Clouds in the far south.

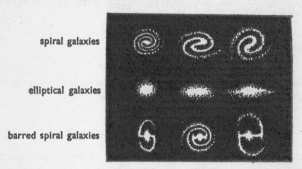

spiral galaxies

elliptical galaxies

barred spiral galaxies

Fig. 30. *Types of galaxies*

Galaxies are of various shapes (Fig. 30). Many are spiral in form, and look like starry Catherine-wheels when observed (or, better, photographed) with large telescopes; others are elliptical or spherical, and a relative few are totally irregular. It used to be thought that the different shapes indicated different stages of evolution, so that, for instance, a spiral galaxy would change gradually into an elliptical, but this is now regarded as dubious. We cannot yet pretend to know much about the development of the galaxies.

Hubble's system of classification is still in use. There are spirals of different degrees of 'tightness'; barred spirals, in which the arms come from a central bar; ellipticals, ranging from Eo (to all intents and purposes globular) to E7 (extremely flattened); and irregulars. The *Galaxy in which we live appears to be a typical spiral.

Apart from over two dozen relatively close systems making up the so-called *Local Group, all the galaxies appear to be receding from us; this is shown by the *red shifts in their spectra, and has led to the general conclusion that the whole universe is expanding. The further away a galaxy lies from us, the faster it is receding.

As well as sending out visible light, some galaxies are very powerful radio sources; the cause of this intense radio emission is not yet properly understood. It was once thought that some such sources were due to colliding galaxies – that is to say, galaxies 'passing through' each other in opposite directions, but this idea has now been rejected.

Galaxies tend to occur in clusters – there are, for instance, major clusters seen in the constellations of Coma and Hydra. Yet only the comparatively near galaxies may be studied in detail. We have learned that there are objects such as star-clusters, gaseous nebulæ, and novæ and supernovæ in them, but much remains to be learned.

It must be admitted that apart from the Magellanic clouds, the galaxies are very unspectacular objects in small telescopes. Even the Andromeda Spiral appears as nothing more than a dim patch of haze. Photographs taken with giant instruments are needed to bring out the true forms.

Galaxy, The. The system of stars of which our Sun is a member. It contains about 100,000 million stars, arranged in a shape which has been likened to that of two fried eggs clapped together back to back (Fig. 31)! There are spiral arms, and the Sun, with its planets, lies on the inside of a spiral arm. The diameter of the Galaxy is about 100,000 light-years, while the maximum breadth is perhaps 20,000 light-years. The Sun lies about 32,000 light-years from the centre; the actual nucleus of the Galaxy cannot be seen, because there is too much interstellar material in the way, but the star-clouds in the constellation of Sagittarius (the Archer) indicate its direction.

Fig. 31. *Shape of the Galaxy – 'side' and 'face' views*

Radio astronomy studies have allowed astronomers to plot the positions of the spiral arms, though with limited accuracy. It has also been found that the Galaxy is rotating around its centre; the Sun takes some 225,000,000 years to complete one revolution – a period which has been unofficially termed the *cosmic year*.

It is also known that the Galaxy is by no means exceptional. Of its companions in the local group, one, the Andromeda Spiral, is considerably larger.

Galilei, Galileo (1564–1642). Italian scientist, who was the first great telescopic observer. His championship of the theory of a central Sun, rather than a central Earth, led to his condemnation by the Inquisition in Rome. Galileo was also the true founder of experimental mechanics.

Galle, Johann (1812–1910). German astronomer. He and H. D'Arrest were the first to identify *Neptune, in 1846, on the basis of calculations by *Le Verrier.

Gamma-Rays. Extremely short-wavelength electromagnetic radiations.

Ganymede. The third and largest satellite of Jupiter. It has a diameter of 3270 miles according to very recent measurements, so that it is larger than the planet Mercury. It is so bright that it would be an easy naked-eye object were it not so close to Jupiter in the sky, and binoculars show it excellently.

Ganymede has recently been studied by radar, the equipment used being the 210-foot antenna at Goldstone in California. It seems that the most likely possibility for the surface is rocky and/or metallic material embedded in a matrix of ice; such a surface could be relatively smooth, with a top layer of ice rubble, but it might be rough, since to radar the ice would be transparent. Craters could not be seen, probably because Ganymede was so far away during the radar test (380,000,000 miles) but a photograph of the satellite from Pioneer 11 in December 1974 showed apparent craters as well as two maria.

Gauss. The standard unit of measurement for a magnetic field. The Earth's magnetic field, at the surface, ranges between 0·3 and 0·6 gauss.

Gauss, Karl F. (1777–1855). German mathematician – one of the greatest in scientific history. His calculations led to the rediscovery of the minor planet *Ceres a year after Piazzi had found it, in 1901.

Gegenschein. A very faint glow in the sky, exactly opposite to the Sun. It is excessively difficult to observe, and from Britain is seldom seen; in size, it may cover an area equal to that of the Square of Pegasus. Its origin is uncertain, but it is presumably due to thinly-spread material in space, and it is thought to be associated with the *Zodiacal Light. The Gegenschein is sometimes called by its English name of *Counterglow.

Geiger Counter. An instrument used to detect charged particles and high-energy radiation. Most artificial satellites and space-probes carry Geiger counters.

Geocentric Theory. The old plan of the *Solar System, according to which the Earth lay in the central position with the Sun, Moon and planets moving round it.

Geodesy. The study of the figure, dimensions, mass and other associated features of the Earth.

Gibbous Phase. The phase of the *Moon when between half and full (Fig. 32). *Mercury and *Venus, of course, show lunar-type phases, and *Mars may also appear decidedly gibbous at times.

Fig. 32. *Gibbous Phase*

Globular Clusters. See *Clusters, Star.

Globules. Small dark patches inside gaseous *nebulæ. It is thought possible that they may be stars in the process of formation. Globules are best displayed in the famous *Orion Nebula.

Gnomon. The 'pointer' of a *sundial; its function is to cast its shadow on the dial in order to give the time. The gnomon must always be pointed toward the celestial pole.

Goodricke, John (1764–1786). English astronomer, born in Holland. He discovered the variability of *Delta Cephei, and explained the cause of the variations of *Algol. Goodricke was a deaf-mute; despite this, he would have had a brilliant career if he had lived.

Gould's Belt. A belt of bright stars, inclined to the *Milky Way at about 20 degrees. It was first pointed out by Sir John Herschel in the 19th century, and was studied by the American astronomer B. Gould; it includes most of the bright stars in Orion, Scorpio, Carina and Centaurus as well as some bright stars in other constellations. It seems to be due to the fact that near the Sun, there is a slight 'tilting' of the spiral arm of the *Galaxy in which the Sun lies. The effect is most noticeable for the bright, naked-eye stars of spectral type B.

Granules, Solar. The Sun's bright surface or *photosphere is not smooth; when seen in detail, a granular structure appears, each granule being several hundreds of miles in diameter. The granules are short-lived, lasting for only a few minutes, and are in constant motion. The best photographs of them so far obtained have been taken by instruments carried in high-altitude balloons.

Gravitation. The force of attraction which exists between all particles of matter in the universe. Its fundamental nature is still quite unknown.

Great Circle. A circle on the surface of a sphere (such as the *Earth, of the *celestial sphere) whose plane passes through the centre of the sphere. Thus a great circle will divide the sphere into two equal parts. The horizon, the celestial equator and the ecliptic are all great circles on the celestial sphere (Fig. 33).

Greek Alphabet. See *Stars.

Green Flash (or Green Ray). When the Sun is on the point of setting, its last visible portion may flash brilliant green for a moment. This is due to effects of the Earth's atmosphere, and is

Fig. 33. *Great circles on the Earth's surface*

best seen over a sea horizon. Venus has also been known to show a green flash when setting.

Greenwich Mean Time (G.M.T.). The time at Greenwich, reckoned according to the mean Sun. It is used as the standard throughout the world.

Greenwich Meridian. The line of longitude which passes through Greenwich Observatory (or, to be more precise, a special instrument at Greenwich known as the Airy *transit circle). It is still taken as longitude 0 degrees, even though the official Greenwich Observatory has now been moved to Herstmonceux in Sussex.

Greenwich Observatory. The most famous British observatory, founded in 1675. Until the retirement of Sir Richard Woolley, in 1971, the Director of the Observatory was also Astronomer Royal. Greenwich is still regarded as the timekeeping centre of the world, and the zero for longitude is marked by the transit circle installed there by *Airy. Deteriorating conditions led to the transfer of the equipment to *Herstmonceux, in Sussex, and the original Greenwich Observatory is now a museum.

Gregorian Calendar. The *calendar now in use.

Gregorian Reflector. A type of reflecting telescope in which the light from the object under study strikes the main mirror, and is reflected back up the tube on to a small concave mirror placed outside the focus of the main mirror (Fig. 34). The light then

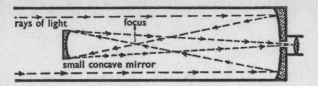

Fig. 34. *Principle of the Gregorian reflector*

comes back through a hole in the main mirror, as with the
*Cassegrain, and reaches the eyepiece. Telescopes of this kind
were first described by James Gregory, three centuries ago, and
were popular for a while; but they are not easy to make or adjust,
and few of them are now in use. A Gregorian, unlike a *Cassegrain or *Newtonian reflector, gives an erect image (north at
the top, south at the bottom).

H

H-I and H-II Regions. Clouds of hydrogen in the *Galaxy. In H-I regions this hydrogen is neutral (that is to say, each atom is complete) and the clouds cannot be seen, though they can be detected by radio astronomy methods. In H-II regions, the hydrogen is ionized (that is to say, each atom is incomplete), and the presence of hot stars makes the cloud shine as a *nebula. A really hot star may have an effect stretching out to as much as 500 light-years from it.

Halation Ring. A ring sometimes seen round a star image on a photograph. It is, of course, purely a photographic effect.

Hale, George Ellery (1868–1938). American astronomer, noted for his solar work; he invented the *spectrohelioscope. He was also mainly responsible for the setting-up of great telescopes, the last of which was the 200-inch at Palomar.

Hale Observatories. This is the name now given to the two great American observatories at Mount Wilson and Palomar, in California. The Mount Wilson has a 100-inch reflector; the Palomar 200-inch reflector was until very recently the world's largest. It was completed in 1948, and has been responsible for many fundamental advances in astronomy.

Halley, Edmond (1656–1742). The second Astronomer Royal. He made an expedition to St Helena to study the southern stars; he undertook fundamental researches into the motion of the Moon; he financed the publication of Newton's *Principia*, and he was the first to predict the return of a comet (*Halley's Comet), though the return did not take place until 1758, long after his death.

Halley's Comet. The only bright *comet to have a period of less than a century. It takes 76 years to complete one journey round the Sun, and its regular visits have been recorded back to the year BC 240. It is next due in 1986 (Fig. 35). The comet is

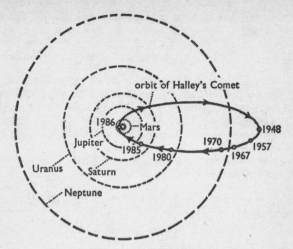

Fig. 35. *Orbit of Halley's Comet*

named in honour of Edmond *Halley, the second Astronomer Royal, who was the first to realize that it appears periodically.

Halo. The spherical-shaped star cloud round the main part of the *Galaxy. It is best referred to as the *galactic halo*, to distinguish it from the meteorological halo (a luminous ring seen round the Sun or Moon due to ice crystals in the Earth's atmosphere).

Harvest Moon. In the northern hemisphere, the full moon nearest the autumnal *equinox is termed Harvest Moon; the autumnal equinox occurs on September 22 or thereabouts. In the ordinary way, the Moon rises more than half an hour later from one night to the next; this time-lapse is known as the *retardation*. It may sometimes amount to over an hour, so that if the Moon rises at, say, 22 hours G.M.T. on one evening it may not rise until 23 hours G.M.T. on the next evening. Near the autumnal equinox, however, the retardation is no more than 15 minutes – though it is incorrect to say, as many books do, that the Moon then rises at almost the same time for several nights in succession. It is also wrong to suppose that the Harvest Moon looks larger than any other full moon.

Heliacal Rising. The rising of a celestial body at the same time as the rising of the Sun. The more common use of the term is, however, 'the date when a celestial body first becomes visible in the dawn'. Thus Sirius is said to rise heliacally when it can first be made out in the dawn sky, in the late autumn. The ancient Egyptians paid great attention to the heliacal rising of Sirius.

Heliocentric Theory. The theory according to which the Sun lies at the centre of the *Solar System, as proposed by Copernicus – and, very much earlier, by the Greek philosopher Aristarchus. The word comes from the Greek *helios* (=Sun).

Heliometer. A refracting telescope in which the object-glass is cut in half, so that one half may be made to slide past the other – giving a double image of the object under study. It is used to measure very small distances as seen through the telescope eyepiece, so that its purpose is the same as that of a *micrometer even though its principle is quite different.

Hellas. A prominent circular feature on *Mars, once thought to be a plateau, but now known to be a depressed basin. It lies to the south of the *Syrtis Major, and can at times be very bright.

Henderson, Thomas (1798–1844). Scottish astronomer. While Director of the Cape Observatory, in the 1830s, he measured the *parallax of Alpha Centauri, but did not work out the star's distance until after *Bessel had announced the distance of the star 61 Cygni.

Hermes. A very small *asteroid, only a mile or two in diameter. In 1937 it passed within 485,000 miles of the Earth – so far, the closest known celestial body apart from the Moon.

Herschel, Caroline (1750–1848). Sister and colleague of *William Herschel. She was herself an able observer, who discovered six comets.

Herschel, Sir John (1792–1871). Son of *William Herschel. He too was a great observer; from 1833 to 1838 he was at the Cape of Good Hope, making the first systematic studies of the southern stars.

Herschel, Sir William (1738–1822). Probably the greatest of all astronomical observers. He discovered *Uranus in 1781 with a home-made telescope, but his main work was in connection with the distribution of the stars. He discovered the binary nature of some double stars, as well as many *clusters and *nebulæ.

Herschelian Reflector. A type of reflecting telescope developed by *Sir William Herschel in the late 18th century. The main mirror is tilted, and the light is brought to focus at the side of the upper end of the tube, thus removing the need for any secondary mirror (Fig. 36). However, there are numerous disadvantages to this arrangement, and Herschelian reflectors are now seldom or never to be found.

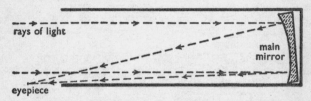

Fig. 36. *Principle of the Herschelian reflector*

Herstmonceux. The present observing site of the Royal Greenwich Observatory. It is in Sussex, near the town of Hailsham. The present Director is Dr F. Graham Smith. The largest telescope has been the 98-inch reflector, but this is now to be moved to a new site out of England.

Hertzsprung, Ejnar (1873–1967). Danish astrophysicist; one-time Director of the Leiden Observatory in Holland. In 1905 he discovered the giant and dwarf divisions of the stars.

Hertzsprung-Russell Diagram. A diagram in which stars are plotted according to their spectra and their absolute *magnitude (Fig. 37). As is described under the heading *Star, the stars are divided into various types; O, B and A (white), F and G (yellow), K (orange) and M (red; the remaining spectral types, R, N and S, are also red, but are relatively rare).

When the diagram is drawn, as was done before the first world war by E. J. Hertzsprung and H. N. Russell, a definite pattern emerges. Most of the stars lie on a well-defined belt known as the *Main Sequence, but there are some very large and luminous stars of spectrum M and K making up a *giant branch*, while near the lower left-hand portion of the diagram are the very small, dense *White Dwarfs. The Sun is a typical main sequence star of spectrum G.

Hertzsprung-Russell or 'H-R' diagrams have been of great value in theoretical astrophysics. It was once thought that a star began as a giant, shrank and heated up until joining the Main Sequence at the upper left (type O or B) and then passed down the Main Sequence, ending its career as a red dwarf of spectrum M; but it is now known that the situation is much more complicated than this, and for much of its career a star remains at much the same point on the Main Sequence – after which it moves on to the giant branch. Instead of being youthful, a giant star is now thought to be nearing the end of its brilliant career, since it has used up much of its nuclear 'fuel'.

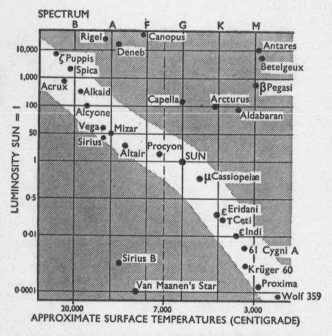

Fig. 37. *Hertzsprung-Russell Diagram*

Hevelius, Johann (Hewelcke) (1611–1687). Danzig astronomer, who drew up a notable star catalogue and a lunar map.

Himalia. The sixth satellite of Jupiter.

Hipparchus (BC 190–120). Greek astronomer, who drew up a star catalogue; *Ptolemy's catalogue was based upon it. Hipparchus also discovered *precession. Little is known of his life, and his original catalogue has not come down to us.

Horizon. The great circle on the *celestial sphere which is everywhere 90 degrees from the observer's overhead point or *zenith. The actual horizon will not usually be quite the same as the apparent horizon, partly because of irregularities in ground-level and partly because of the height of the observer himself.

Horrocks, Jeremiah (1619–1641). English astronomer, who made the first observations of a *transit of Venus (1639). His early death cut short a career of exceptional promise.

Hour Angle. The time which has passed since a celestial object crossed the *meridian. (The meridian is the great circle on the *celestial sphere which passes through the zenith and the two poles of the sky, so that it cuts the horizon at the north and south points.) If the object has not yet crossed the meridian, and is therefore in the eastern part of the sky, its hour angle is negative. To find the hour angle of a body, simply subtract its *right ascension from the local *sidereal time.

Hour Circle. A great circle on the *celestial sphere passing through both the north and south poles of the sky. The zero hour circle coincides with the observer's meridian.

Hubble, Edwin (1889–1953). American astronomer, who worked extensively with the Mount Wilson 100-inch reflector. He made many notable contributions, but is best remembered for his work in connection with the distances of galaxies. In 1923 he discovered Cepheids in some of the external systems, and was able to show that they lie far beyond our Galaxy. This was almost certainly the most important discovery since the time of Newton.

Hubble's Constant. Spectroscopic work has shown that the universe is expanding, and that the more distant *galaxies are receding from us more quickly than closer ones. The rule is: the further, the faster. The increase of velocity with distance seems to follow a definite law, first calculated by the American astronomer E. E. *Hubble, and known therefore as Hubble's Constant.

Huggins, Sir William (1824–1910). English amateur astronomer; one of the great pioneers of stellar spectroscopy.

Hunter's Moon. The full moon following *Harvest Moon. Since Harvest Moon usually occurs in the last part of September, in the Northern Hemisphere, Hunter's moon is seen in October.

Huygens, Christiaan (1629–1695). Dutch scientist; one of the ablest observers of his time. He was the first to recognize the nature of Saturn's rings, and he discovered *Titan. Huygens also made the first successful pendulum clock, and improved telescope designs.

Hyades. The open star *cluster near Aldebaran. Aldebaran is not, in fact, a member of the cluster; it lies in the foreground.

Hyperion. The seventh *satellite of Saturn.

I

Iapetus. The eighth *satellite of Saturn. It is smaller than the Moon, and is much brighter when west of Saturn than when to the east of the planet, so that either it is irregular in shape or else has an unequally reflecting surface.

Icarus. The only *asteroid known to have its perihelion closer to the Sun than the orbit of Mercury.

Imbrium, Mare. Vast lunar 'sea', visible with the naked eye.

Inferior Conjunction. When the planets *Mercury and *Venus lie almost between the Earth and the Sun – that is to say, their right ascension is the same as that of the Sun – they are said to be at inferior conjunction; they are, of course, 'new', and so cannot be seen unless the lining-up is exact enough to produce a *transit. Only bodies which are closer to the Sun than we are can reach inferior conjunction – that is to say Mercury, Venus, some comets, and a few minor planets.

Inferior Planet. The planets *Mercury and *Venus, which have orbits inside that of the Earth.

Infra-red Radiations. Radiations with wavelength longer than that of red light (7500 *Ångströms), and which cannot be seen with the eye. The infra-red region extends up to the short-wave end of the radio part of the *electromagnetic spectrum.

Interferometer, Radio. See *Radio Telescopes.

Interferometer, Stellar. An instrument for measuring the diameters of stars. It depends upon the principle of light-interference.

International Geophysical Year (I.G.Y.). The period between July 1957 and the end of December 1958, when scientists of over fifty nations co-operated in a programme of observations to learn more about all aspects of the Earth. The data obtained during the I.G.Y. took many years to analyse, and a tremendous amount of new knowledge was gained.

Interstellar Matter. Material which is spread out between the stars of a *galaxy. It consists largely of hydrogen together with minute solid particles known generally as 'dust'. The average density is incredibly low – no more than one atom of hydrogen per cubic centimetre. Gaseous *nebulæ may be regarded as condensations of interstellar material, and there is similar material also between the stars of other galaxies, such as the *Andromeda Spiral.

Io. The first satellite of Jupiter. Information from the Pioneer probes shows that it has a tenuous ionosphere, and it seems to influence the Jovian radio emission. It is slightly larger than our Moon. See *Satellites.

Ion. An *atom which has lost one or more of its *electrons, and so has a positive electrical charge, since in a complete atom the positive charge of the nucleus is balanced out by the combined negative charge of the electrons. The process of producing an ion is known as *ionization*. An atom which has gained extra electrons is *positively ionized*.

Ionosphere. The region of the Earth's atmosphere above the *stratosphere. It extends between about 40 and about 500 miles above the ground, and contains the layers which reflect some radio waves back to Earth, thus making long-range wireless communication possible.

Irradiation. The effect which makes brilliantly-lit or self-luminous bodies appear larger than they really are.

Isaac Newton Telescope. The 98-inch reflecting telescope, set up at *Herstmonceux but now being moved to the new *Northern Hemisphere Observatory.

J

Janus. Innermost satellite of Saturn, discovered by Dollfus in 1966. It is very small, and can be seen only when the rings are edge-on. See *Satellites.

Jeans, Sir James (1877–1946). Famous British astronomer; noted for his work in cosmology and astrophysics as well as for his popular books and broadcasts.

Jodrell Bank. The site, near Macclesfield in Cheshire, of Britain's leading radio astronomy observatory. The most famous instrument is the 250-foot. paraboloid. The present Director is Sir Bernard Lovell.

Jones, Sir Harold Spencer (1890–1960). The tenth Astronomer Royal. He made many notable contributions, and was also in office when – partly at his instigation – the equipment of the Royal *Greenwich Observatory was moved to a better site at *Herstmonceux.

Julian Day. A count of the days, reckoning from 12 noon on January 1, BC 4713 – a starting point chosen quite arbitrarily by the mathematician Scaliger, who introduced the system in the year 1582. The 'Julian' is in honour of Scaliger's father, whose Christian name was Julius, and has nothing to do with Julius Caesar. Julian Day reckoning is often used in variable star recording. January 1 1966 was Julian Day 2,439,126.

Juno. The third member of the swarm of *Minor Planets. It was discovered by K. Harding in 1804, and has a diameter of about 150 miles. It never becomes bright enough to be visible with the naked eye.

Jupiter. The largest planet in the *Solar System (Fig. 38). Its mean distance from the Sun is 483,000,000 miles, and its revolution period is 11·9 years. Details of its orbit and dimensions are given under the heading *Planets.

Because Jupiter is so large, with a diameter of 88,700 miles as measured through the equator, it is very brilliant; it is outshone only by *Venus and (occasionally) by *Mars. Moreover, Jupiter is well placed for observation for several months in each year. During the late 1970s it will be in the northern hemisphere of the sky. *Oppositions will occur in October 1975 (in Aquarius), November 1976 (in Pisces), December 1977 (in Aries) and January 1979 (in Cancer).

Through a telescope, Jupiter shows a yellowish, decidedly flattened disk, crossed by the famous cloud belts. The flattening is real; Jupiter's polar diameter is only 83,300 miles. This is because of the planet's quick rotation, which makes the equatorial regions bulge out. The flattening is much greater than with the Earth, where the difference between the equatorial and the polar diameters is only 26 miles – partly because Jupiter is spinning more quickly, and partly because its outer layers, at least, are made up of gas rather than solid rock.

Jupiter does not rotate as a solid body would do. The equatorial region (known as System I) has a period of about 9 hours 50 minutes, and the rest of the planet (System II) about 9 hours 55 minutes, though special features of the disk have rotation periods of their own and so drift about in longitude. Among these is the famous Great Red Spot, an oval area which can be as much as 30,000 miles long by 7,000 miles wide, and which has been under observation for several centuries. Sometimes it vanishes for a while, but it always comes back, whereas the other spots on Jupiter are much shorter-lived. The surface features are always changing.

Jupiter   Earth

Fig. 38. *Comparative sizes of Jupiter and the Earth*

When the outer gases were first analysed, by means of the *spectroscope, it was found that there was a tremendous amount of hydrogen, together with hydrogen compounds such as methane (marsh-gas) and ammonia. It was once believed that there must be a rocky core, surrounded by a thick layer of ice which was in turn overlaid by the deep, hydrogen-rich atmosphere; alternatively, it was suggested that hydrogen might be the main

78

constituent throughout the globe, though near the core it would be so compressed that it would start to assume the characteristics of a metal. In 1955, too, it was found that Jupiter is a source of radio waves, and the presence of a strong magnetic field appeared probable.

Telescopic studies of Jupiter are of particular interest. Generally there are two very prominent belts, the North Equatorial and the South Equatorial; the South Temperate Belt can also be very strong, while usually the other belts are weaker. Fig. 39 shows the main belts and zones.

In December 1973 the first Jupiter probe, Pioneer 10, by-passed the planet at a distance of about 80,000 miles, and sent back pictures of excellent quality, together with a vast amount of miscellaneous information. The intense magnetic field was confirmed, and it was also found that there were zones of very strong radiation - which very nearly put the instruments on the probe out of action. Pioneer 11 followed in December 1974, and it too was a complete success; this time the vehicle approached the planet from the south pole, swung quickly across the dangerous equatorial region where the radiations were at their strongest, and then went across the north pole before leaving Jupiter behind. The north pole was covered by a relatively featureless hood, while the equatorial region showed convective cells. There seems to be a drop in cloud level toward the poles.

Jupiter has an immense, spinning magnetosphere which is constantly buffeted by the *solar wind, and is also affected by the inner large satellites. The magnetic field itself is of a very complicated nature, and it is also known that Jupiter periodically sends into space great numbers of high-energy electrons. The planet is indeed extremely active.

Following the first pass of Jupiter, Pioneer 10 began a never-ending journey into space - carrying a plaque, in the admittedly vain hope that some alien civilization will eventually find it. Pioneer 11 was sent onward toward a rendezvous with *Saturn; this should take place in 1979. By then a Mariner probe will be well on its way, and should pass Jupiter not so very long after Pioneer 11 has done so.

It is fair to say that Jupiter is much the most important body in the Solar System apart from the Sun, and it must have a high internal temperature, perhaps reaching half a million degrees. It is also sending out more energy than might be expected, and this could be explained by a very slight shrinking, so that the excess energy would be gravitational; but there is no suggestion that Jupiter is self-luminous, and the outer clouds are extremely

79

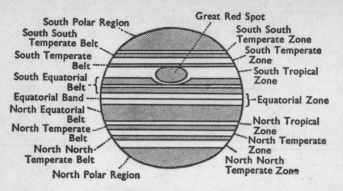

Fig. 39. *Belts and zones of Jupiter*

cold. Below the clouds there must be a region where the temperature is comparable with that of the Earth, but if life of any kind exists there it must be very different from anything with which we are familiar, and many people (including myself!) are decidedly sceptical about it.

According to a recent theory which has met with wide acceptance, Jupiter has a small rocky core, made up of iron and silicates at about 30,000°. Above this is a very deep layer of liquid; an inner shell, out to 30,000 miles from the centre of the planet, made up of liquid metallic hydrogen, surrounded by a layer of liquid molecular hydrogen; and finally the Jovian atmosphere. On this theory (due to J. D. Anderson and W. B. Hubbard) the excess energy is not due to shrinking or radioactivity, but merely to the remnant of the original heat.

Jupiter has fourteen satellites. Four (Io, Europa, Ganymede and Callisto) are bright enough to be seen with any small telescope; the remaining ten are very small. Details are given under the heading *Satellites.

K

Kepler, Johannes (1571–1630). Great German mathematician, who used the observations made by Tycho *Brahe to show that the planets move round the Sun not in circles, but in ellipses.

Kepler's Laws. Three important laws of planetary motion, announced by the German mathematician J. Kepler between 1609 and 1618. They are as follows:
1. The planets move in elliptical orbits, the Sun being situated at one focus of the ellipse.
2. The radius vector, or imaginary line joining the centre of the planet to the centre of the Sun, sweeps out equal areas in equal times. (In other words, a planet moves at its fastest when it is at its closest to the Sun.)

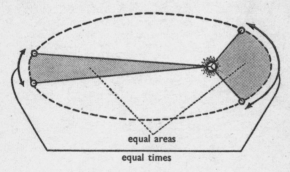

equal areas

equal times

Fig. 40. *Kepler's Second Law*

3. The squares of the *sidereal periods of the planets are proportional to the cubes of their mean distance from the Sun.
It was the announcement of these laws, following years of careful work by Kepler, which really disproved the old *geocentric or Ptolemaic theory of the universe, and showed that Copernicus had been right in saying that the planets move round the Sun. Kepler's Laws also apply, of course, to bodies such as comets, and to satellites moving round planets (Fig. 40).

Kiloparsec. One thousand *parsecs (3260 *light-years).

Kirkwood Gaps. Regions in the belt of *Minor Planets in which almost no minor planets are found.

Most of the minor planets have orbits lying wholly between those of Mars and Jupiter; but a body whose period is a simple fraction of that of Jupiter (one-quarter or one-half, for instance) is repeatedly influenced or *perturbed* by Jupiter's powerful gravitational pull until it has been forced into a different path. The result is that the minor planets tend to avoid orbits which would give them any exact fraction of Jupiter's own $11\frac{3}{4}$-year period. These 'empty' regions are known as Kirkwood Gaps, in honour of the last-century American mathematician Daniel Kirkwood, who first drew attention to them.

Kohoutek's Comet. A comet discovered in 1973 by the Czech astronomer L. Kohoutek. It was expected to become brilliant at the end of 1973, but failed to do so. It has an estimated period of about 75,000 years.

Kuiper, Gerard P. Dutch-American astronomer; one of the greatest of planetary observers and a pioneer of space-research methods. He died in 1973. A crater on the planet *Mercury has been named in his honour.

L

Lagrange, J. L. (1736–1813). Great French mathematician (born in Italy) who was one of the pioneers of modern dynamical astronomy.

Laplace, P. S. (1749–1827). French mathematician, best remembered for his *Nebular Hypothesis, published in 1796.

Lassell, William (1799–1880). English amateur astronomer, noted for his planetary work. He was prevented by a sprained ankle from joining in the hunt for *Neptune, in 1846; but seventeen days after Neptune had been located, Lassell discovered its larger satellite, *Triton.

Latitude, Celestial. The angular distance of a celestial body from the nearest point on the *ecliptic.

Leavitt, Henrietta Swan (1888–1921). American woman astronomer. She discovered 2400 variable stars, 4 novæ and several minor planets, but is best remembered for her discovery, in 1912, of the *Period-Luminosity Law of the Cepheids.

Leda. The thirteenth satellite of Jupiter.

Leiden Observatory. The leading astronomical observatory of Holland.

Leonids. The November *meteor shower. In most years the Leonids are sparse, but occasionally they give brilliant displays, as in 1799, 1933, 1866 and 1966. They are associated with a faint periodical comet, Tempel's.

Le Verrier, U. J. J. (1811–1877). Great French astronomer, whose calculations led to the discovery of *Neptune.

Libration. Though the Moon always keeps the same face turned toward the Earth, because its rotation period is the same as the time taken for it to complete one journey round us (or, more accurately, round the *barycentre), we can examine a total of 59% of the surface at various times; the remaining 41% is permanently averted. The fact that from Earth we can see more than half the surface, though never more than 50% at any one moment is due to effects known as librations.

Though the Moon rotates on its axis at a constant rate, it does not move in its orbit at a constant speed, because its path is not circular – and, following *Kepler's Law, it moves at its quickest when it is at its closest to us. This means that the rotation and the position in orbit must become periodically 'out of step', so to speak, and the Moon seems to oscillate slightly; first a little of the western limb is exposed, and then, some days later, a little of the eastern limb. This is *libration in longitude*. There is also a *libration in latitude*, due to the fact that the Moon's equator is tilted to the plane of its orbit by over 6 degrees; and thirdly there is a daily or *diurnal libration*, because the Earth is itself spinning, taking the observer with it. When the Moon is on the horizon, the observer is 'elevated' above the centre of the Earth by about 4000 miles (the Earth's radius) so that he can see for an extra one degree round the edge of the Moon.

Lick Observatory. Major American observatory, on Mount Hamilton, California. Its main instrument is a 120-inch reflector, but it also contains the world's largest refractor, the 40-inch made by *Clark and completed in 1888.

Life on Other Worlds. Nobody is sure how life on Earth began, but at any rate we can lay down some conditions which are necessary for a world bearing life of our kind. There must be an atmosphere, containing sufficient oxygen; there must be water, and there must be a suitable temperature-range. In the Solar System, only the Earth seems to fulfil all these requirements.

Most of the planets and satellites may be ruled out at once. There is no trace of life, either past or present, on the Moon; the rocks which have been brought back from there are completely sterile. *Mercury has virtually no atmosphere, and *Venus is as hostile as it could be; the surface is fiercely hot, and the atmosphere is made up chiefly of carbon dioxide. The giant planets have no visible solid surfaces, and the suggestion that life might exist beneath the outer clouds of *Jupiter seems to be distinctly far-fetched, though some astronomers consider that it is not out of the question. *Pluto, of course, is hopelessly cold, and there is no evidence of atmosphere there. Among the satellites, only *Titan, the senior attendant of Saturn, has a reasonably dense atmosphere; and Titan is unsuitable in other ways.

84

We are left, then, with *Mars. Here there is a thin atmosphere (the ground pressure is much less than with Titan) and the climate is chill by our standards; it now seems definite that the dark surface patches are not vegetation-tracts, as used to be thought before the era of space-probes. It is possible that lowly organic matter may survive, but we will not know for certain until the first soft landings with automatic vehicles take place. These are scheduled for 1976.

Yet it must always be remembered that the *Sun is an ordinary star, and few modern astronomers doubt that other stars have planet-systems of their own. There must surely be millions upon millions of 'other Earths' capable of supporting life, and we may assume that intelligence is spread widely through the universe, though we cannot tell what form it may take. The only way to establish contact seems to be by means of radio. If we could pick up radio signals of a regular, rhythmical pattern, we could at least infer that the transmissions must be artificial. In 1960 the American radio astronomers at Green Bank, West Virginia, made a systematic search for such signals, but – not surprisingly! – the famous Project *'Ozma' was fruitless, and was soon discontinued.

The detection of *pulsars, in 1969, caused a flurry of excitement. The signals were so regular that for a few days astronomers concerned in the discovery seriously wondered whether the pulses were non-natural; but this intriguing idea – since nicknamed the LGM or Little Green Men theory – was soon found to be wrong. We now know that pulsars are *neutron stars.

One thing is certain: sending a rocket probe out toward the stars, with a view to establishing contact, is pointless. At speeds we can manage at the moment, a journey even to the nearest star would take many thousands of years. Pioneer 10, the Jupiter probe which by-passed the Giant Planet in late 1973, is now on its way out of the Solar System, and carries a plaque which could tell any alien civilization the system from which the probe was launched; but the chances of Pioneer ever being found are negligible. And if we are to achieve interstellar travel, it must be by some method about which we know nothing as yet – so that speculation is both endless and futile.

Of course, it is always possible that there are alien life-forms which can exist under conditions which are totally unsuitable for Earthmen. Story-tellers make great play of what are popularly known as BEMs or Bug-Eyed Monsters; but in sober fact, the scientific information we have at present argues against the existence of alien life anywhere in the universe.

Light-Curve. A graph showing the changing brightness of a *variable star. The magnitude of the star is plotted against the period, as shown in the diagram (Fig. 41). In some cases the light-curve is regular, as with *Cepheids, while for other variable stars the curve is quite irregular.

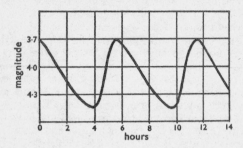

Fig. 41. *Light-curve of δ Cephei*

Light-Year. The distance travelled by light in one year. Since light moves at 186,000 miles per second, a light-year is equal to 5,880,000,000,000 miles, or rather less than six million million miles.

Linné. Small feature on the lunar Mare Serenitatis. In 1866 the German astronomer Schmidt announced that Linné had changed from a crater into a white patch, but few astronomers now believe in any change there. The modern form of Linné, as photographed from *Orbiter probes, is that of a small, well-marked crater.

Local Group. A group of *galaxies, of which the three most important members are the *Andromeda Spiral, the Triangulum Spiral, and our own *Galaxy, with the two *Nubeculæ or Clouds of Magellan. There are over two dozen fainter members. The galaxies in the Local Group are not receding from us, though all galaxies beyond the group are racing away at high velocities.

Lockyer, Sir Norman (1823–1920). Pioneer English astrophysicist and spectroscopist. He founded the Norman Lockyer Observatory near Sidmouth, in Devon.

Longitude, Celestial. The angular distance of a body from the

86

ernal *equinox or *First Point of Aries, measured eastward
along the *ecliptic from 0 degrees to 360 degrees.

Lowell Observatory. The observatory at Flagstaff, in Arizona,
founded by *Percival Lowell. It is noted for its planetary work,
and it was here, in 1930, that *Pluto was discovered.

Lowell, Percival (1855–1916). American astronomer, best remembered for his now-rejected theories of artificial canals on
Mars. He was a skilled mathematician, whose calculations led
to the discovery of the planet *Pluto.

Lunation. The interval between one new moon and the next
(or one full moon and the next). It is equal to 29 days 12 hours
44 minutes. An alternative name for it is *synodical month*.

Lyot, Bernard (1897–1952). Eminent French astronomer. He
invented the *coronagraph and the *Lyot filter. All his career
was spent at the *Meudon Observatory, near Paris.

Lyot Filter. A device used for observing the Sun's *prominences,
and other features of the solar atmosphere, without waiting for
an *eclipse. It may also be called a *monochromatic filter*.

Lysithea. The tenth satellite of Jupiter.

M

Mädler, J. H. (1794–1874). German astronomer, who, with *W. Beer, drew up the best lunar map of its time.

Magellanic Clouds (or Clouds of Magellan). See *Nubeculæ.

Magnetic Storm. A sudden disturbance of the Earth's magnetic field, shown by interference with radio communications as well as by variations in the compass needle. It is due to charged particles sent out by the Sun, usually associated with solar *flares.

Magnetosphere. The region of the Earth's magnetic field, which is of considerable extent, and includes the *Van Allen zones. It resembles a teardrop in shape, with the tail pointing away from the Sun. On the sunward side of the Earth, the magnetosphere extends out to about 40,000 miles, but on the night side of the Earth it reaches to a much greater distance. As the *solar wind comes toward the Earth it meets the magnetic field, producing a shock-wave; inside the shock-wave there is a turbulent region, inside which is a definite boundary, the magnetopause. The magnetosphere proper lies on the Earthward side of the magneto-pause. On the dark side of the Earth, the shock-wave merely weakens until it can no longer be detected.

The *Moon's magnetic field is virtually nil, while that of Mercury is somewhat stronger, giving rise to an appreciable magnetosphere; the same applies to *Venus, and almost certainly to *Mars as well. *Jupiter has a very strong magnetic field, and zones of intense radiation (which very nearly put the instruments of Pioneer 10 out of action permanently). So far as we know at the moment, *Saturn does not have a comparable field, while about the remaining planets we have no information.

Magnitude. This is really a term for 'brightness', but astronomically there are several different kinds of magnitudes.

Apparent or *visual magnitude* is the apparent brightness of a celestial body. The brighter the object, the lower the magnitude; thus the brilliant star Aldebaran in Taurus is of magnitude 1, while the faintest stars normally visible to the naked eye are of magnitude 6. The brightest stars have zero or, in a few cases, negative magnitudes; that of Sirius, the most brilliant star in the sky, is −1·4. On the other hand, the world's largest telescope can show stars down to as faint as magnitude +23. The magnitude scale is very precise; a star of the first magnitude is one hundred times as bright as a star of magnitude 6. It is important to note that a star's apparent magnitude is no reliable key to its real luminosity; thus the Pole Star (magnitude 2·0) is more than three magnitudes fainter than Sirius, but it is also much more remote, and is very much more luminous.

On the stellar scale, the brightest planet, Venus, has a magnitude of about −4·6, while the full moon reaches −12, and the Sun −27.

Absolute magnitude is the apparent magnitude that a star would have if it were seen from a distance of 10 *parsecs, or 32·6 *light-years. At this distance, Sirius would have a value of +1·3, whereas the Pole Star would be a brilliant object of magnitude −4·6. Absolute magnitude is, therefore, a measure of the star's real luminosity. The absolute magnitude of our Sun is +4·8, so that from a distance of 10 parsecs it would be a dim object as seen with the naked eye.

Photographic magnitude has been described under the heading *Colour Index.

Ordinary magnitudes are measured according to the amount of visible light received, but with a *bolometric magnitude* all the other wavelengths of the *electromagnetic spectrum are included as well, so that the bolometric magnitude is a measure of the total amount of radiation being sent out by the star. The value for the Sun is +4·6.

Main Sequence. If the stars are plotted on a graph of the sort known as a *Hertzsprung-Russell Diagram, according to luminosity and spectral type, it will be found that most of them lie on a well-marked band, beginning with hot white stars (spectra O and B) and ending with feeble red stars (spectrum M). This is the Main Sequence. The Sun, a yellow dwarf of spectral type G, is a typical Main Sequence star.

Maksutov Telescope. A special type of astronomical telescope, making use of both mirrors and lenses. It gives excellent results, and has a wide field of view, but is not easy to make. It was first described in detail in 1944 by the Russian astronomer after whom it is named.

Mars. The fourth planet in order of distance from the Sun.

Details of its orbit and globe are given under the heading *Planets.

Mars was named after the mythological God of War because of its strong red colour. When at its closest to the Earth (rather less than 35 million miles) it may outshine any celestial body apart from the Sun, the Moon and Venus, but generally it is less bright than Jupiter, and when badly placed it may look like nothing more than a fairly conspicuous red star not much superior to Polaris. It comes to *opposition (Fig. 42) only once in about 780 days, so that there are long periods when it is too far away to be well seen. The next oppositions will be those of December 1975 (in Taurus), January 1978 (in Gemini/Cancer) and February 1980 (in Leo). Because the orbit of Mars is of appreciable eccentricity, not all oppositions are equally close, as Fig. 42 shows. The opposition of 1971 took place with Mars near *perihelion, and was therefore close (less than 35,000,000 miles); that of 1980 will take place with Mars near *aphelion (distance over 60,000,000 miles).

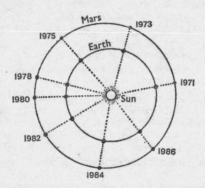

Fig. 42. *Oppositions of Mars, 1971–1986*

Mars has always been regarded as the most Earthlike of the planets, despite its relatively small size; its diameter is 4200 miles, roughly intermediate between those of the Earth and the Moon. The escape velocity of 3·2 miles per second at once indicates that the atmosphere is likely to be thin; indeed, it is now known to be much more tenuous than used to be thought before the age of space-probes. It has also been found that the main atmospheric constituent is carbon dioxide, and the ground pressure is nowhere as high as 10 millibars.

The poles of Mars are covered with whitish caps, which wax and wane according to the seasons; during Martian winter a cap may be very large and prominent, though during summer it may virtually disappear. It is now believed that the caps are made up chiefly of solid carbon dioxide, though there may also be a certain amount of ordinary ice. They were first seen by observers during the 17th century, and can often be really striking.

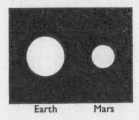

Earth   Mars

Fig. 43. *Comparative sizes of the Earth and Mars*

The dark markings were discovered by Christiaan Huygens in 1659, and have been carefully mapped; they are essentially permanent, though variations in tone and extent do occur. The most prominent of them is the V-shaped Syrtis Major (formerly known as the Kaiser Sea or the Hourglass Sea), which is visible with a small telescope when Mars is well placed. Originally it was thought that the dark regions were seas, and the red areas deserts; but when it became clear that Mars cannot have large sheets of water on its surface, the theory generally accepted was that the dark patches were 'vegetation' – presumably growing in old sea-beds. Before the flight of Mariner 4, in 1965, most astronomers inclined to the view that lowly organisms existed on Mars.

The situation has since been transformed by the Mariner probes, of which the most successful up to the present time has been Mariner 9. The vehicle reached the neighbourhood of Mars in November 1971, and for the next few months sent back spectacular pictures of high resolution, showing not only many craters – previously recorded by Mariner 4 in 1965 and Mariners 6 and 7 in 1969 – but also huge volcanoes, together with features which look remarkably like dry riverbeds. The loftiest of the volcanoes, such as the Olympia Mons and the Arsia Mons, tower to about fifteen miles above the general surface level, so that they are higher and more massive than any volcanoes on Earth.

91

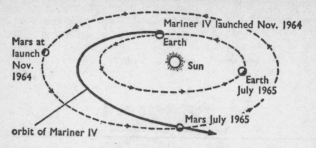

Fig. 44. *Path of Mariner IV*

Major dust-storms can take place on Mars, as in 1971 and 1973, but rainfall is certainly unknown at the present epoch. There is, however, one major mystery. The 'riverbeds' and the great systems of canyons give the impression of having been water-cut, and they are not seriously eroded, so that by geological standards they can hardly be very ancient; it may be that a few tens of thousands of years ago Mars had more water and atmosphere than it has now, and it has even been suggested that this state of affairs may recur in the future. At present, however, we can rule out the possibility of any advanced life, and it is quite likely that the planet is completely sterile. We may hope to find out in the summer of 1976, when the two American Viking probes are scheduled to soft-land on the surface and send back information. The Mariners have shown that the famous 'canals' – straight, artificial-looking lines criss-crossing the planet, drawn by Percival Lowell and his colleagues in the early part of our own century – do not exist, and there is no trace whatsoever of intelligent activity on Mars. Yet it is true to say that Mars is less hostile than any other world in the Solar System apart from Earth, and there is no reason to doubt that colonies will be established there within the next hundred years or so.

Mars has two small satellites, Phobos and Deimos, both discovered by Asaph Hall, at Washington, in 1877. Mariner photographs have shown that they are irregular in shape, and that they are pitted with craters. Both are very small (less than twenty miles in diameter), and are quite unlike our Moon; they may well be ex-asteroids which were captured by Mars in the remote past.

Maskelyne, Nevil (1732–1811). The fifth Astronomer Royal. In 1767 he founded the Nautical Almanac.

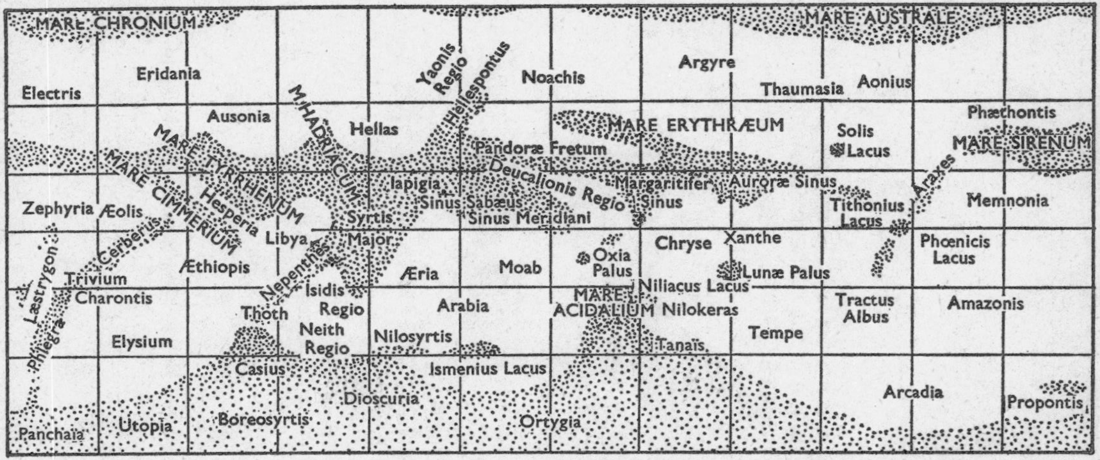

Fig. 45. *Map of Mars*

Mass. The quantity of matter that a body contains. It is no the same as weight; for instance, an astronaut on the Moon has only one-sixth of his Earth weight, because the Moon has a much weaker pull of gravity, but his mass is unaltered.

McDonald Observatory. Major American observatory on Mount Locke (7000 ft. high) in Texas. It has a great 82-inch reflector.

McMath-Hulbert Observatory. Solar observatory in Michigan, USA.

Mean Sun. An imaginary body travelling eastward along the celestial *equator, at a rate of motion equal to the average rate of the real Sun along the *ecliptic. Further details are given under the heading *Equation of Time.

Megaparsec. One million *parsecs.

Earth   Mercury

Fig. 46. *Sizes of Earth and Mercury compared*

Mercury. The closest to the Sun of the principal planets. It is never prominent as seen with the naked eye, partly because it is small – its diameter is only 3000 miles – and partly because it remains in the same part of the sky as the Sun. When at its best, it may show up as a bright star either low in the west after sunset or low in the east before sunrise. Because it is nearer the Sun than we are, it shows *phases (Fig. 47) similar to those of Venus, so that it may be seen as a crescent, half or *gibbous shape; when full it is on the far side of the Sun, and is virtually out of view. Occasionally it may pass directly between the Sun and the Earth, appearing in *transit as a black disk against the bright face of the Sun. The revolution period, or Mercurian 'year', is 88 Earth-days.

Very little detail can be seen on Mercury, even with large telescopes, and before 1974 our knowledge of the surface was decidedly meagre. However, in March 1974 and again in September the probe Mariner 10 by-passed the planet and sent back excellent pictures, showing that the surface is as rough and crater-scarred as that of the Moon. There is practically no atmosphere, but there is an appreciable magnetic field.

Mercury is a slow spinner; its rotation period is 58·7 Earth-days, so that to an observer at a fixed position on Mercury the interval between one sunrise and the next would be 176 Earth-days or two Mercurian years. The temperature range is very great, and there seems to be no chance that any life exists there. Like Venus, Mercury has no satellite.

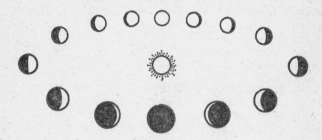

Fig. 47. *Phases of Mercury and Venus*

Meridian, Celestial. The great circle on the *celestial sphere which passes through the *zenith and both celestial poles. It follows that the meridian cuts the observer's horizon at the exact north and south points. The *upper meridian* is the section from the south point up through the zenith to the north celestial pole. (This applies, of course, to the northern hemisphere of the Earth; from countries such as Australia, the upper meridian will extend from the north point up through the zenith to the south celestial pole.)

Messier, Charles (1730–1817). French astronomer, interested mainly in comets but best remembered for his catalogue of nebulæ and star-clusters. See *Messier numbers.

Messier Numbers. In the latter part of the 18th century, Charles Messier drew up a catalogue of nebulous objects, consisting of star-*clusters, gaseous *nebulæ and *galaxies. The list contained 102 objects, to which 5 were added later. Messier's numbers are still widely used – thus the Andromeda Galaxy is M.31, the Orion Nebula M.42, and the famous star-cluster Præsepe, in Cancer, M.44 – but in 1886 the catalogue was to all intents and purposes superseded by the much more accurate and extensive New General Catalogue of Clusters and Nebulæ, drawn up by the Danish astronomer J. Dreyer, then Director of the Armagh Observatory in Northern Ireland.

Messier himself was not particularly interested in nebulæ. He spent his time in hunting for new comets, and found that he often mistook a nebula or cluster for a comet – so he listed all nebulous objects in order to be able to avoid them. Ironically, his catalogue is still widely used, though very few people now remember much about the comets that he discovered during his career!

Meteor. A small particle, usually smaller than a grain of sand, moving round the Sun in the manner of a dwarf planet. When moving freely in space it cannot be seen, because it is too small, but when it dashes into the Earth's upper atmosphere, moving at a relative speed of anything up to 45 miles per second, it rubs against the air-particles and becomes heated by friction, destroying itself and producing the luminous effect known as a *shooting-star*.

It is thought that over 20,000,000 meteors enter the atmosphere each day, but each is of very slight mass, and few penetrate below a height of 50 miles above the ground without being destroyed. Their origin is not known. They may be the fragments of a larger planet (or planets) that met with some disaster in the remote past; on the other hand, they may be nothing more than the débris left over when the main members of the Solar System were formed.

Meteors tend to travel in swarms. Each time the Earth passes through a swarm, the result will be a shower of shooting-stars; the meteors of a shower will seem to radiate from one particular point in the sky, known as the *radiant, but this is merely an effect of perspective, as a little thought will show. A man standing on a bridge overlooking a motorway will see the traffic lanes apparently radiating from a point in the distance; the lanes are really parallel, and the principle is the same with shower meteors, which move through space in parallel paths.

The most spectacular shower of each year takes place in August, and is known as the Perseid shower, since the radiant lies in the constellation Perseus. Much less reliable are the November Leonids, which are sometimes spectacular – as in 1833, 1866 and

1966, but in other years may be very feeble. The most important annual meteor showers are as follows:

January 4.	Quadrantids.	Radiant in Boötes. Sharp maximum.
April 19–22.	Lyrids.	Moderate shower.
July 27–August 17.	Perseids.	Rich, reliable shower.
October 15–25.	Orionids.	Moderate shower.
November 14–17.	Leonids.	Rich occasionally.
December 9–13.	Geminids.	Quite rich.

In addition to these showers, there are *sporadic meteors*, not connected with any particular showers, which may appear from any direction at any moment. They are much less frequent than shower meteors.

It is now thought that all meteors are true members of the Solar System, and do not come from interstellar space. It is also known that there is a close association between meteors and *comets.

Meteorite. A relatively large body which survives the complete drop to Earth without being destroyed. There is a fundamental difference between a meteor and a meteorite; it now seems that meteorites are very near relations of *minor planets, and there may be no distinction between a large meteorite and a small minor planet.

Meteorites are divided into several classes, according to their chemical composition. Stony objects are known as *aerolites*; iron meteorites are termed *siderites*, and stony-irons as *siderolites*. Each class is again sub-divided, according to the elements that it contains. When etched with acid, a siderite that has been cut and polished shows characteristic patterns known as *Widmanstätten patterns.

The largest known meteorite is still lying where it fell, in prehistoric times, at Hoba West in Africa; it weighs over 60 tons. Another meteorite which landed in Arizona in the remote past has left a large crater, almost a mile in diameter, while there are other meteorite craters in Australia and elsewhere. During the present century there have been only two major falls, one in 1908 and the other in 1947, both in Siberia – though it must be added that some Russian astronomers believe the 1908 object to have been the nucleus of a small *comet.

The most famous British meteorite of recent years fell at Bnrwell, in Leicestershire, on Christmas Eve 1965. It flashed across the Midlands, attracting considerable attention, and many fragments of it were found. The original weight of the meteorite may have been about 200 pounds. One fragment went through the window of a house in the village, and was discovered later in a vase of artificial flowers!

Most museums have meteorite collections. The average meteorite is however small, weighing only a few pounds, and large ones are extremely rare. There is no record of anyone having been killed by a falling meteorite.

Meudon Observatory. Major French observatory, near Paris. It contains the famous 33-inch refractor as well as much other equipment.

Micrometeorites. Extremely small particles, no more than 1/250 of an inch in diameter, moving round the Sun. Their masses are too low for them to produce luminous effects when they enter our atmosphere, and therefore they do not cause shooting-star appearances. Since 1957 they have been carefully studied by means of artificial satellites and space-probes.

Micrometer. A measuring instrument. used together with a telescope to measure very small distances – such as the separations of the components of double stars. There are various forms, the most common being the *filar micrometer.

Micron. A unit of length, equal to 1/1000 of a millimetre (1/25,400 of an inch); there are 10,000 *Ångströms to one micron. The usual symbol for a micron is the Greek letter μ (mu).

Midnight Sun. The Sun seen above the horizon at midnight. This can occur during some part of the year anywhere inside the Arctic or Antarctic Circles. At the North Pole, the Sun stays above the horizon all the time that it is north of the celestial *equator – that is to say, between the time of the spring *equinox (about March 21) and the autumnal equinox (about September 22). Throughout this period, of course, the Sun remains below the horizon at the South Pole.

Milky Way. The luminous band stretching across the night sky. It has been known since the earliest times, and there are many legends about it, but it was not until the invention of the telescope that it was found to be made up of large numbers of faint stars. As explained under the heading *Galaxy, it is due to a line-of-sight effect, and the stars in the Milky Way are not genuinely crowded together.

The term used to be applied to the Galaxy itself, but nowadays the name Milky Way is restricted to the luminous band seen in the sky.

Mimas. The first satellite of Saturn (though now, since the discovery of Janus (1966), second in order of distance from the planet). See *Satellites.

Minor Planets. Between the orbits of Mars and Jupiter lie many thousands of small bodies, known officially as Minor Planets but also called *asteroids* or *planetoids* (Fig. 48). The first and largest of them, *Ceres, was discovered in 1801 by G. Piazzi; with its diameter of perhaps 700 miles, it is much the largest of the whole swarm. Apart from *Vesta, which is on the limit of naked-eye visibility when best placed, no minor planet is bright enough to be seen without optical aid.

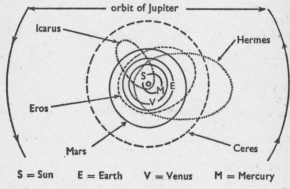

S = Sun   E = Earth   V = Venus   M = Mercury

Fig. 48. *Orbits of exceptional asteroids*

Minor planets are too small to retain atmospheres, and many of them are probably not regular in shape. Their origin is uncertain. It has been suggested that they, together with the meteoric bodies in the Solar System, represent the fragments of an old planet which broke up for some reason or other; but it is thought more possible that they are nothing more than material scattered during the formation of the principal planets.

Details about the first ten minor planets to be discovered may be given in a table, as follows:

Number	Name	Sidereal period (years)	Mean distance from the Sun (miles)	Mean opposition magnitude
1	Ceres	4·60	257,000,000	7·4
2	Pallas	4·61	257,400,000	8·7
3	Juno	4·36	247,800,000	8·0
4	Vesta	3·63	219,300,000	6·0
5	Astraea	4·14	239,300,000	9·9
6	Hebe	3·78	225,200,000	8·5
7	Iris	3·68	221,500,000	8·7
8	Flora	3·27	204,400,000	9·0
9	Metis	3·69	221,700,000	8·3
10	Hygeia	5·59	292,600,000	9·5

Some of the minor planets have orbits which take them away from the main swarm, and may bring them relatively close to the Earth. Of these, the best-known is *Eros, discovered in 1898 and which may come within 15,000,000 miles of the Earth as it did in early 1975. The present holder of the 'approach record' is *Hermes, which by-passed us in 1937 at less than twice the distance of the Moon, but which is only about one mile in diameter. Another dwarf minor planet, Icarus, has an orbit which takes it closer to the Sun than Mercury; on the other hand the minor planets of the *'Trojan' group move round the Sun at the same mean distance as Jupiter, while the remarkable Hidalgo has a very eccentric orbit which swings it out almost as far as Saturn.

The total number of minor planets is certainly very great. R. S. Richardson, the American astronomer, has estimated 44,000, while some Russian astronomers believe 100,000 to be nearer the truth.

Minute of Arc (symbol: ′). One-sixtieth of a *degree of arc.

Miranda. The innermost *satellite of Uranus. It is too faint to be seen with telescopes of amateur size.

Molecule. A stable association of atoms – or, broadly speaking, a group of atoms linked together. A molecule of water, for example, is made up of two hydrogen atoms together with one oxygen atom (H_2O), while a molecule of the gas carbon dioxide consists of one carbon atom together with two oxygen atoms (CO_2). Some molecules may be extremely complex.

Monochromatic Filter. An alternative name for the *Lyot filter.

Month. In everyday language, a month – more accurately, a *calendar month* – is either 30 or 31 days, apart from February, which has 28 days in ordinary years and 29 days in each leap-year. Astronomically, however, there are several kinds of 'months', all associated with the time taken for the Moon to complete one journey in its orbit.

The *anomalistic month* of 27·55 days is the time taken for the Moon to travel from one *perigee to the next.

The *sidereal month* (27·32 days) is the time taken for the Moon to go once round the Earth – or, to be precise, round the *barycentre – with reference to the stars.

The *synodical month* or *lunation* (29·53 days) is the time between successive new moons, or successive full moons.

The *nodical* or *Draconitic month* (27·21 days) is the time taken for the Moon to make successive passages through one of its *nodes.

The *tropical month* (27·32 days) is the time taken for the Moon to return to the same celestial longitude. It is only about 7 seconds shorter than the sidereal month.

Moon. The Moon is generally called the Earth's satellite. However, it has a mass of 1/81 of that of the Earth, which makes it rather too considerable to be a mere satellite (Fig. 49); it is better to regard the Earth-Moon system as a double planet.

Earth   Moon

Fig. 49. *Relative sizes of Earth and Moon*

It used to be thought that the Moon broke away from the Earth in the remote past, leaving a depression now filled by the Pacific Ocean, but there are strong mathematical objections to this attractive idea, and it has now been given up. Modern astronomers believe that the two worlds have always been separate.

The Earth and Moon move round the *barycentre, or centre

of gravity of the Earth-Moon system, but since the barycentre lies within the Earth's globe it is good enough, for most purposes, to say that the Moon revolves round the Earth in a period of 27·32 days. Because it has no light of its own, and shines only by reflected sunlight, it shows regular *phases from new to crescent, half, gibbous, full and back again to new. Its orbit is somewhat eccentric (Fig. 50), so that its distance from the Earth ranges between 221,460 miles at *perigee to 252,700 miles at *apogee. The real diameter of the Moon is 2160 miles; the apparent diameter varies between 29′ 22″ and 33′ 31″.

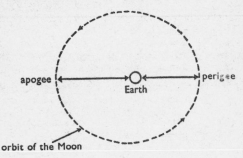

Fig. 50. *Lunar perigee and apogee*

Though the Moon appears so brilliant in the night sky, it sends us only 1/465,000 of the light of the Sun. Moreover, it has a low *albedo, and reflects only 7% of the sunlight which falls upon its surface.

The Moon's gravitational pull is much less than that of the Earth, and a man standing on the lunar surface would have only 1/6 of his Earth weight. This feeble gravity means that the Moon has lost almost all of any atmosphere it may once have had, so that it is now what is commonly called an 'airless world'. The surface temperatures are extreme, ranging, on the lunar equator, between about +220 degrees Fahrenheit at noon to below −250 degrees Fahrenheit at midnight. Since the rotation period is the same as the sidereal period (27·32 days), a lunar 'day' is almost as long as a fortnight on Earth.

The Moon's surface (Fig. 51) contains broad, darkish plains, still known as 'seas' (or, in Latin, *mares*) even though there is no water in them – and almost certainly never has been, though they must originally have been 'seas' of lava. There are high peaks, and vast numbers of walled formations known as craters,

Fig. 51. *Outline chart of the Moon*

ranging from enclosures well over 100 miles in diameter down to tiny pits too small to be seen from Earth.

Because the axial rotation period is the same as the sidereal period, the Moon keeps the same face turned toward us all the time, though the various *libration effects mean that we can examine a total of 59% of the surface instead of only 50%. The first photographs of the averted hemisphere were sent back in 1959, from the Russian probe Luna 3. The American Orbiters of the mid-1960s obtained detailed pictures, and by now we have excellent maps of virtually the whole of the Moon. The main difference between the two hemispheres is that the far side contains no major maria.

There have been endless discussions as to the origin of the lunar walled formations. According to one theory, the craters are of internal origin ('volcanic', using the term in a very broad sense), while other astronomers attribute them to *meteorite impact. No doubt both processes have operated. Even now the problem has not been solved, despite the Apollo landings between 1969 and 1972.

As everyone knows, the first men to reach the Moon were Neil Armstrong and Edwin Aldrin, in July 1969. Apollo 11 marked the beginning of a new era, and since then there have been five further landings. Lunar samples have been brought home for analysis, both by the American astronauts and by Russian automatic probes, so that our knowledge of the Moon has been improved out of all recognition. The rocks are essentially basaltic and similar to those of the Earth, though somewhat different in detail, and it also seems that the Moon and the Earth are of about the same age (between 4·5 and 5 thousand million years). The atmosphere is negligible, and so is the overall magnetic field. Moreover, the Moon is – and presumably always has been – sterile. Minor ground tremors take place, and are associated with the elusive phenomena known as *T.L.P.s (Transient Lunar Phenomena), but any major activity on the lunar world belongs to the remote past.

Fig. 52. *Cross-section of lunar crater*

Recording instruments set up on the Moon by the Apollo teams are still active, and information continues to be sent back, but it seems that the next landings will be deferred for at least ten years – perhaps longer. However, it is likely that full-scale lunar research bases will be set up before the end of the century, and these will be of tremendous value to all mankind.

Moon Illusion. When the Moon is low down over the horizon, it seems to look larger than when it is high up. This is a pure illusion, since in reality the low Moon is no larger than the high Moon. Ptolemy, the last great astronomer of ancient times, wrote that an object seen across 'filled space' (such as the Moon near the horizon, seen across the Earth's surface) will give the impression of being more distant than an object seen across 'empty space' (such as the high Moon), so that it will seem to be larger. This may not be the full explanation, but in any case, it is easy to prove, by practical experiment, that the illusion *is* an illusion and nothing more.

Mount Wilson Observatory. American observatory, established in 1904 due to the efforts of G. E. *Hale. Among its instruments is the 100-inch Hooker reflector, for many years the largest telescope in the world.

Multiple Star. A star made up of more than two components. A famous example is Castor, in the constellation of Gemini (the Twins). To the naked eye it appears as a rather bright star; a telescope shows that there are two components; with the spectroscope, it is possible to find that each component is itself a close *binary, while there is a fainter companion, some way away, which also is a close binary. Castor therefore consists of six stars, four bright and two dim. Another well-known multiple is Theta Orionis, the so-called 'Trapezium' in the *Orion Nebula.

Mural Quadrant. This is described under the heading *Quadrant.

N

Nadir. The point on the celestial sphere immediately below the observer. Obviously the nadir is directly opposite to the overhead point or *zenith.

Nasmyth, James (1808–1890). British engineer; a pioneer lunar observer, and also the inventor of the steam-hammer.

Nebula. A mass of tenuous gas in space, together with what is loosely termed 'dust'. If there are stars contained in a nebula, the gas and dust will become visible, either because of straightforward reflection or because the stellar radiation causes the nebular material to emit light on its own account. If there are no suitable stars, the nebula will remain dark, and will be detectable only because it blots out the light of stars beyond.

The most famous of the bright nebulæ is the Sword of Orion (M.42), which is easily visible to the naked eye. It contains hot early-type stars, and is an *emission nebula*; that is to say, it gives out some light of its own. On the other hand, the Pleiades star-cluster contains a *reflection nebula*, difficult to see visually even with large telescopes, and well shown only on long-exposure photographs. Many *dark nebulæ* are known. The most conspicuous of them, the 'Coal Sack' in the Southern Cross, can never be seen from Europe, but there is another well-marked dark nebula in the constellation of Cygnus (the Swan).

Nebulæ are of great interest, because it is thought that they are the birth-places of stars. Thousands of millions of years ago our own Sun was probably born inside a nebula, by condensing out of the tremendously rarefied nebular material, and it is believed that similar processes are going on all the time; the strange *globules may well be stars in the process of formation.

Gas-and-dust or *galactic nebulæ* are common enough, though few of them are visible with the naked eye. Similar objects are found in other galaxies, such as the Andromeda Spiral and the Clouds of Magellan. Incidentally, the old terms 'spiral nebula' and 'extragalactic nebula' have now gone out of fashion, and external systems are known simply as *galaxies.

We must also note the so-called *planetary nebulæ, which will be described separately. They are not true nebulæ, but merely stars with extended gaseous 'atmospheres'.

Nebular Hypothesis. An old theory of the origin of the Solar System, put forward in 1796 by the French mathematician Laplace. According to Laplace, the planets were formed from 'rings' thrown off by a shrinking cloud of gas; the Sun was supposed to be the last remnant of the cloud.

The Nebular Hypothesis was widely accepted for many years, but it failed to stand up to mathematical analysis, and had to be rejected. However, modern theories such as that due to Von Weizsäcker follow roughly the same principles. The question is discussed in slightly more detail under the heading *Solar System.

Neptune. The eighth planet in order of distance from the Sun. Details of its orbit and dimensions are given under the heading *Planets.

Neptune was discovered in 1846. Uranus, which had been found in 1781, had been moving in an apparently erratic fashion, and it had refused to follow its predicted path, so that presumably some unknown body was pulling it out of position. Two mathematicians, U. J. J. Le Verrier in France and J. C. Adams in England, independently worked out where this new body must be. Their calculations were remarkably accurate, and the planet was soon identified by J. Galle and H. D'Arrest, at the Berlin Observatory, working upon Le Verrier's calculated position.

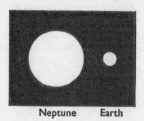

Neptune   Earth

Fig. 53. *Comparative sizes of Neptune and Earth*

Neptune is well below naked-eye visibility (magnitude 7·7), but a small telescope will show it. It is a giant planet 31,200 miles in diameter (Fig. 53), but it never approaches us within 2,600,000,000 miles, so that surface details on it are hard to see even with the world's largest telescopes. Its outer parts, like those of the other giants, are made up of gas.

Neptune has two satellites. One of them, Triton, is perhaps

3000 miles across, so that it may be seen with a moderate instrument; the other, Nereid, is extremely small and faint.

Nereid. The smaller of Neptune's two *satellites; it is a very faint, elusive object.

Neutrino. A fundamental particle. Since it has no mass and no electric charge, it is far from easy to detect.

Neutron. Neutrons are particles which exist in the nuclei of all *atoms apart from the hydrogen atom. Each neutron has a mass about equal to that of a *proton, but it has no electric charge.

Neutron Star. The remnant of a very massive star which has suffered a *supernova outburst. Following the collapse of the core, the protons and electrons have run together, making up neutrons, and the density is amazing – it has been estimated that a cubic inch of neutron star material would weigh over 4000 million tons. The diameter of the star may be between 5 and 100 miles; there may be a thin solid crust in which sudden disturbances or 'starquakes' take place.

Neutron stars send out rapidly-varying radio waves, and are therefore called *pulsars. According to current theory, there is a very strong magnetic field; the star is rotating quickly, and if the magnetic pole does not coincide with the pole of rotation there is a kind of searchlight-beam effect. Every time we pass through the 'beam' we receive a radio pulse. The only neutron star which has so far been optically identified with certainty is the pulsar in the *Crab Nebula. Further details are given under the heading *Pulsars.

Newcomb, Simon (1835–1909). American mathematical astronomer; a leading authority on celestial mechanics.

Newton, Sir Isaac (1642–1727). Greatest of all mathematicians. His *Principia*, published in 1687, introduced the 'modern' era of astronomy.

Newtonian Reflector. The most common type of reflecting telescope. It is named after Sir Isaac Newton, who developed it and who built the first such reflector in or about 1671.

With a Newtonian, the light from the object to be studied passes down an open tube on to a curved mirror. This mirror sends the light back up the tube on to a smaller mirror, or *flat*, placed at an angle of 45 degrees. The light is then directed into the side of the tube, where an image is formed and magnified by an *eyepiece (Fig. 54). In a Newtonian reflector, therefore, the observer looks into the tube instead of up it. A certain amount of light is lost because the flat gets in the way, but this cannot be helped, and the loss is not serious.

Newtonians are very convenient. The mirrors, of course, need periodical attention; the glass surface is covered with a thin layer of silver, aluminium or rhodium, which has to be renewed when it becomes dull or patchy, but silvering at least is not difficult. The telescope tube may be of skeleton construction, and in some Newtonians a *prism is used instead of a flat secondary mirror.

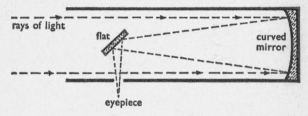

Fig. 54. *Principle of the Newtonian reflector*

Nix Olympica. (Old name for Olympus Mons.) Great Martian volcano, about 15 miles high.

Noctilucent Clouds. Rare, strange clouds in the *ionosphere, best seen at night when they continue to catch the rays of the Sun. They are over 50 miles high, and are quite different from ordinary clouds. It is possible that they are produced by dust left by meteors which have burned away in the upper air.

Nodes. The points at which the orbit of the Moon, a planet or a comet cuts the plane of the *ecliptic (Fig. 55). When the body crosses the plane of the ecliptic as it moves from south to north, it is said to pass the *ascending node*; when the movement is from north to south, the body passes the *descending node*. The line joining these two points is called the *line of nodes*.

North Polar Distance. The angular distance of a celestial body from the north celestial *pole.

North Polar Sequence. A list of 96 stars, near the north pole

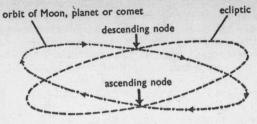

Fig. 55. *Nodes*

of the sky, whose photographic *magnitudes have been measured as accurately as possible. Other stars are then photographed and compared with the stars in the North Polar Sequence, so that their own magnitudes may be worked out. The Sequence includes stars from magnitude 2 (Polaris) down to as faint as magnitude 20.

Northern Hemisphere Observatory. A proposed new observatory to be used by British and other astronomers. The main telescopes will be an 180-inch reflector; the 98-inch *Isaac Newton reflector, transferred from *Herstmonceux; and a 40-inch or 50-inch reflector.

Nova. A star which suffers a sudden outburst, and flares up to many times its normal brightness for a few days, weeks or months before fading back to its former obscurity The word 'nova' is Latin for 'new', but a nova is not truly a new star, as used to be thought.

Novæ are not due to collisions between two stars. This plausible-sounding idea was disproved long ago. Stellar collisions must be extremely rare even in the most crowded parts of the Galaxy, whereas novæ are not particularly uncommon, even though relatively few become visible with the naked eye. A nova outburst occurs in the star itself, though generally the outburst is confined to the star's outer layers. A few stars such as T Coronæ have been known to undergo more than one outburst, and are termed *recurrent novæ.

Probably the most interesting nova of modern times was that in Delphinus (HR Delphini), discovered in 1967 by the English amateur astronomer G. E. D. Alcock. It remained visible with the naked eye for months, and at the end of 1975 was still within the range of a small telescope. Its pre-nova magnitude was 12, so that in all probability it will not fade much below

110

this level. Alcock discovered a nova in Vulpecula in the following year, and he has since found a telescopic nova in the constellation of Scutum.

The following are the brightest novæ to have been seen since the turn of the century:

Year seen	Constellation	Maximum magnitude
1901	Perseus	0·0
1912	Gemini	3·3
1918	Aquila	−1·1
1920	Cygnus	2·0
1925	Pictor	1·1 (Too far south to be seen from Europe)
1934	Hercules	1·2
1936	Lacerta	1·9
1942	Puppis	0·4 (Too low to be well seen from Europe)
1963	Hercules	3·2
1967	Delphinus	3·7
1968	Vulpecula	4·9
1970	Serpens	4·4
1975	Cygnus	1·8

Novæ have also been recorded in other galaxies, but their great distances mean that they are very faint.

Nubeculæ (or Magellanic Clouds). The nearest of the external star-systems, and so the brightest as seen with the naked eye; they look rather like broken-off parts of the *Milky Way, and the Large Cloud, or Nubecula Major, is visible even in bright moonlight. Unfortunately for observers in Europe and the United States, the Clouds lie in the far south of the sky, and never rise above the horizon in northern countries.

Both systems are considerably smaller than our Galaxy. The Large Cloud is about 40,000 light-years in diameter, and the Small Cloud 20,000, which is much less than the 100,000 light-years of the system in which we live. They are more or less irregular in form (though traces of spiral structure have been suspected in the Large Cloud), and it has been suggested, though without proof, that they may be 'satellite galaxies' moving round our own. Each is some 180,000 light-years from us, but the distance between their centres is only 75,000 light-years, and radio investigations have shown that both Clouds are contained in a common envelope of very thinly-spread hydrogen gas.

The Clouds are exceptionally important scientifically, because they contain objects of all kinds: open and globular *clusters, gaseous *nebulæ, *supergiant stars, *variables and much else. They give, in fact, an excellent sample of the objects also found in our Galaxy, with the advantage that all the objects in the

Clouds are at approximately the same distance from us. For instance, it was by studying *Cepheids in the Small Cloud that Miss Leavitt, in 1913, made the discovery which led to the recognition of the Cepheid period-luminosity law.

The Large Cloud contains the famous irregular variable S Doradûs, which is equal to one million Suns and is one of the most luminous stars known – and yet is so remote, at its distance of 180,000 light-years, that without optical aid it cannot be seen at all! Also in the Large Cloud is a vast gaseous nebula, nick-named the Tarantula Nebula, which is 800 light-years across, as against only 25 light-years for the famous *Orion Nebula in our own Galaxy.

Since the Clouds lie at less than one-tenth the distance of the next nearest of the large galaxies, the Andromeda and Triangulum Spirals, they may be examined in considerable detail. Without them, our knowledge of other galaxies would be much less extensive than it actually is.

Nutation. A slight, slow 'nodding' of the Earth's axis, due to the fact that the Moon is sometimes above and sometimes below the *ecliptic and therefore does not always pull on the Earth's equatorial bulge in the same direction as the Sun. The result is that the position of the celestial pole seems to 'nod' by about 9 seconds of arc to either side of its mean position, in a period of 18 years 220 days. The effect is superimposed on the more regular shift of the celestial pole caused by *precession.

O

Oberon. The fourth *satellite of Uranus.

Object-Glass. The main lens of a refracting telescope. It is also known as an *objective*.

Objective Prism. A small prism, mounted in front of the telescope object-glass. The effect is to produce a small-scale spectrum of each star in the field of view, so that many stars may be studied with one photographic exposure. Of course, the stellar spectra produced in this way are not detailed, but the method has been found very useful.

Obliquity of the Ecliptic. The angle between the *ecliptic and the celestial *equator; it amounts to 23° 26′ 54″. It may also be defined as the angle by which the Earth's axis is tilted from the perpendicular.

Observatory. The usual meaning of the word is 'a building which houses a telescope', but properly speaking an observatory is a full-scale astronomical research station, containing equipment of all sorts. It has been said that a large professional observatory, such as Palomar in California, is almost a city in itself.

Early observatories had no telescopes, but were fitted with measuring instruments, such as *quadrants, to measure the apparent positions of the stars and other bodies in the sky. There are various remains to be seen at Delhi and elsewhere; the buildings are remarkably elaborate. The last great observatory of pre-telescopic times was set up by Tycho Brahe on the Danish island of Hven, and it was here that Tycho used his quadrants to draw up an accurate star catalogue. He left Hven for good in 1596.

When telescopes were invented, observatories were naturally developed to house them. Early national observatories were those of Paris and Copenhagen. Greenwich Observatory, in England, dates from 1675, and has always been regarded as the timekeeping headquarters of the world, since it was set up by special order of King Charles II so that a new star catalogue could be compiled for the use of British sailors who depended upon astronomical methods of navigation. In Ireland, the

113

observatories at Dunsink (near Dublin) and Armagh were both founded before the end of the 18th century, and many other observatories came into being both in Europe and in America.

Now that large telescopes are used for long-exposure photographs, darkness and clarity of air are essential, which means that observatories have to be situated well away from towns. It is also desirable to build them high up, because the unsteadiness of the Earth's atmosphere is at its worst near ground level. The Pic du Midi Observatory, in the French Pyrénées, is almost 10,000 feet up. Palomar, Mount Wilson and Kitt Peak in California, are not so high as this, but are still well elevated. All these establishments contain several telescopes as well as the main instrument for each observatory (the 200-inch reflector at Palomar, the 100-inch at Mount Wilson, and so on); there are workshops, photographic departments and many other features.

The need for atmospheric clarity has also affected Greenwich Observatory, which has now been shifted to Herstmonceux in Sussex; the original buildings at Greenwich, designed by Sir Christopher Wren, are now used only as museums. The largest European telescope is the 102-inch reflector at the Crimean Astrophysical Observatory, in the USSR.

Few people who have not actually visited an observatory have much idea of what goes on there. A professional astronomer spends very little time in looking through a telescope; his work is carried out almost entirely by photography, and for every hour spent inside a dome there must be many hours spent later in analysis and calculation. There are, of course, some observatories which specialize in various branches of research; for instance Arcetri Observatory, in Italy, is concerned almost solely with the Sun.

Amateur astronomers have their own observatories, housing relatively modest telescopes which are nevertheless large enough to be useful. An amateur observatory may take the form of a dome, a run-off shed, or a shed with a sliding roof.

Occultation. The covering-up of one celestial body by another. Thus the Moon may pass in front of a star or (occasionally) a planet; a planet may occult a star, and there have been cases when one planet has occulted another – for instance, Venus occulted Mars in 1590.

Occultations of stars by the Moon are of some importance. Were the Moon's edge surrounded by a dense atmosphere, the star would flicker and fade for a few seconds before vanishing, but in fact it does not – it snaps out abruptly. One moment it is visible; the next, it has been blotted out by the advancing Moon. This in itself is enough to show that the Moon can have no appreciable atmosphere.

The apparent positions of the stars in the sky are known very precisely, but the exact position of the Moon is not easy to predict with the same accuracy. If therefore, an occultation of a star is timed, the position of the Moon's limb at that moment is also known. This is why occultations of stars have been carefully observed in the past.

Strictly speaking, a solar *eclipse is merely the conventional name for an occultation of the Sun by the Moon.

Ocular. An alternative name for an *eyepiece.

Olbers, Heinrich (1758–1840). German amateur astronomer, who discovered two asteroids: *Pallas (1802) and *Vesta (1807).

Olympus Mons. Martian Volcano formerly called *Nix Olympica.

Opposition. The position of a planet when it is exactly opposite to the Sun in the sky, and so lies due south at midnight (Fig. 56). At opposition, the planet, the Sun and the Earth lie in approximately a straight line, with the Earth in the mid position. Obviously, the *inferior planets, Mercury and Venus, can never come to opposition.

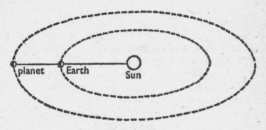

Fig. 56. *Opposition*

Orbit. The path of a celestial object. For instance, the Earth moves round the Sun in a slightly elliptical orbit; *Halley's Comet also moves round the Sun, but in a much more eccentric orbit; *artificial satellites move round the Earth in orbits of various eccentricities and at various distances.

Orion Nebula. The bright gaseous or galactic *nebula in the

Sword of Orion. Its Messier number is 42, but in fact photographs show that both M.42 and the less conspicuous nebula M.43 are concentrations of much more extensive nebulosity covering much of the constellation of Orion.

The Orion Nebula is distinctly visible to the naked eye, below the three bright stars which make up the Hunter's Belt, and moderate telescopes show it well. Immersed in the nebulosity is the celebrated *multiple star Theta Orionis, known as the Trapezium for reasons which will be obvious to anyone who has looked at it.

The distance of the nebula is about 1200 light-years; it is receding at about 11 miles per second. Its density is amazingly low – no more than one-millionth of the best vacuum that can be made in an Earth laboratory. In all probability fresh stars are being formed out of the nebular material, though the process is naturally a slow one. The total diameter of the Orion Nebula is some 25 light-years.

Orrery. A model showing the *Solar System, with the planets capable of being moved at their correct relative velocities round the Sun by some mechanical device (usually by turning a handle). The name comes from Charles Boyle, the fourth Earl of Orrery, who lived during the early 18th century.

Owl Nebula. Messier 97; a planetary nebula in Ursa Major.

Ozma. In 1960 the radio astronomers at Green Bank in West Virginia, under the direction of F. Drake, used the 85-foot radio telescope in a serious search for intelligent life outside the *Solar System. They studied one particular wavelength – 21 centimetres, which is the wavelength of the radiation emitted by clouds of cold hydrogen in the Galaxy – in the hope that they might pick up some signal-pattern which could be shown to be artificial. The experiment was not successful, and was discontinued after a few months, but it was at least worth trying, and it may be repeated in the future. It was officially called Project Ozma, though known more popularly as Project Little Green Men!

P

Pallas. The second largest of the *minor planets. It has a diameter of perhaps 500 miles, but it never becomes bright enough to be visible with the naked eye. It was discovered by H. Olbers in 1802.

Palomar Observatory. Great American observatory in California; it contains the famous 200-inch Hale reflector as well as a 48-inch Schmidt telescope and many other instruments. It and the Mount Wilson Observatory are now known jointly as the *Hale Observatories.

Parallax, Trigonometrical. The apparent shift of a distant body when observed from two different directions (Fig. 57).

The best way to show what is meant is to give a practical experiment. Shut one eye, and line up your finger with an object some way away, such as a clock on the mantelpiece. Now, without moving your finger or your head, open your other eye and shut the first; your finger will no longer be lined up with the clock. If you know the distance between your eyes, and also the angular shift of your finger against the background, it is theoretically possible to work out the distance between your finger and your eyes. Half of the angle of displacement is known as the trigonometrical parallax.

In 1838 the German astronomer F. Bessel applied this method to a star, 61 Cygni. He measured the position of the star twice, with an interval of six months. He was therefore observing the star from opposite sides of the Earth's orbit round the Sun; since the Earth-Sun distance is 93,000,000 miles, he was using a base-line of twice this length, or 186,000,000 miles. He found that 61 Cygni showed a measurable parallax compared with other more remote stars beyond, and Bessel was able to show that it must be about 11 light-years away. Many other stars have since had their distances measured by the same method, but beyond 300 light-years or so the parallax shifts become too small to be detectable.

Parsec. The distance at which a star would show a *parallax of one second of arc. It is equal to 3·26 *light-years, 206,265 *astronomical units, or 19,150,000,000,000 miles. Actually, no

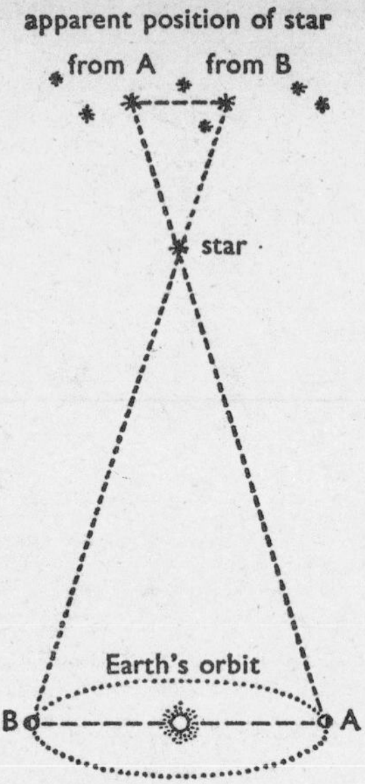

apparent position of star

Fig. 57. *Parallax*

star apart from the Sun is as near as this; the closest star, Proxima Centauri, lies over 4 light-years away from us.

Pasiphaë. The eighth satellite of Jupiter.

P Cygni Stars. Unstable, explosive variable stars. The prototype is P Cygni itself, which for many years now has remained at about the 5th magnitude.

Penumbra. The area of partial shadow to either side of the main cone of shadow cast by the Earth. Its effects are described under the heading *Eclipses, Lunar.

The term is also applied to the outer, relatively light parts of sunspots.

Perigee. This is described under the heading *Perihelion.

Perihelion. The position in the *orbit of a planet or other body when nearest to the Sun (Fig. 58). For instance, the Earth is at perihelion in early January, when the distance between the two bodies is 91½ million miles; at *aphelion, in early July, the distance has increased to 94½ million miles. Similarly, *perigee refers to a body moving round the Earth; the Moon is at perigee when at its closest to us.

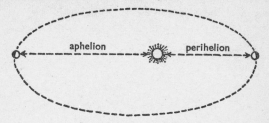

Fig. 58. *Perihelion and aphelion*

Period-Luminosity Law. The relationship between the period and the real luminosity of a *Cepheid variable star.

Perrine, C. D. (1867–1951). Discoverer of eleven comets and two of the satellites of Jupiter (Himalia and Elara). He was for some years Director of the Argentine National Observatory at Cordoba.

Perseids. The August *meteor shower; much the most regular and spectacular of the annual showers.

Perturbations. The disturbances in the orbit of one celestial body produced by the gravitational pulls of others. For instance, the Earth's orbit round the Sun is perturbed by the other planets, notably Venus; *Neptune was tracked down because of its perturbations upon Uranus, and so on. A body of slight mass, such as a *comet, may have its orbit violently perturbed if it passes relatively near a more massive body such as a planet.

Phases. The apparent changes of shape of the Moon, from new to full (Fig. 59).

The Moon has no light of its own, and depends entirely upon reflecting the light of the Sun. When the Moon is almost between the Sun and the Earth, its dark side is turned toward us, and it cannot be seen (though if the alignment is perfect, the result is a solar *eclipse). When the Moon is on the far side of the Earth with respect to the Sun, its lighted half is turned toward us, and the Moon is full; at other times the shape may be a crescent, half, or three-quarter (gibbous) shape.

Mercury and Venus, which are nearer to the Sun than we are, also show complete phases from new to full. Mars may seem decidedly gibbous except when near *opposition, but the other planets are so much further away from the Sun that their phases are too slight to be noticed as seen from Earth.

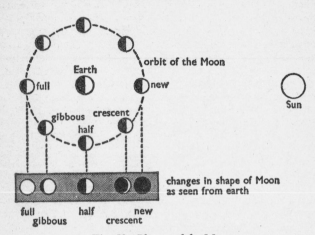

Fig. 59. *Phases of the Moon*

Phillips, T. E. R. (1868–1942). Famous English amateur astronomer, noted for his observations of the planet Jupiter.

Phobos. The inner satellite of Mars. It was photographed from Mariner 9, and found to be an irregularly-shaped body with a longer diameter of less than 20 miles. It has a crater-scarred surface. It is unique inasmuch as its sidereal period (7h 39m) is less than the rotation period of its primary. See *Satellites.

Phœbe. The ninth, outermost *satellite of Saturn. It moves round the planet in a retrograde direction, and may well be a captured *asteroid. It was discovered by Pickering in 1898, but is too faint to be seen with a small telescope.

Photoelectric Cell. An electronic device. Light falls upon the

120

cell, and produces an electric current; the strength of the current depends upon the intensity of the light.

Photoelectric Photometer. An instrument used for measuring the magnitudes of stars or other celestial objects. It consists of a *photoelectric cell used together with a telescope.

Photography, Astronomical. The word 'photography' was coined by an astronomer, Sir John Herschel, in the earlier half of the last century. For more than a hundred years now it has been possible to take good astronomical photographs, and almost all modern research work is carried out in this way, partly because the human eye is notoriously unreliable and partly because it is obviously easier to be able to study pictures in the comfort of a laboratory than to make observations at the eye-end of a telescope. Moreover, the sensitive plate is more efficient than the eye, and it is possible to photograph objects which are too faint to be glimpsed visually even with the world's largest telescopes.

The main trouble about astronomical photography is that time-exposures are needed, amounting often to many hours for faint objects such as remote galaxies. It is quite practicable to guide the telescope accurately while the exposure is being made, but the Earth's atmosphere is always unsteady, as well as being opaque to many of the wavelengths of the *electromagnetic spectrum. However, there have been remarkable improvements in photographic techniques, and excellent results can now be obtained. Almost all work on stellar spectroscopy is also carried out by means of photographs.

Detailed photographs of the Moon are also taken. So far as the planets are concerned, things are less satisfactory, and an observer using a large telescope can see more planetary detail than can be photographed from an Earth-based observatory.

Ordinary astronomical photography can be carried out by amateurs, and spectacular pictures of the Moon, Sun and star-fields can be taken. Even with a simple camera costing a few shillings it is possible to take photographs of star trails; the procedure is to leave the shutter open, so that the stars move across the field from east to west. Trails of meteors and artificial satellites are often recorded during such time-exposures.

The first space photographs of the Moon were obtained in 1959, and since then there have been photographic probes to Mars, Venus, Mercury and Jupiter. Obviously, these pictures have told us more than we could ever have learned in any other way. Beyond the Earth's atmosphere, moreover, the whole of the electromagnetic spectrum is available for study, and important results have already been obtained – in the relatively new science of *X-ray astronomy, for instance.

Photometer. An instrument used to measure the intensity of light coming from one particular light-source. Astronomically, there are several kinds. The old *wedge photometer* used a sliding scale, darkened to different degrees along its length; the star was observed at various points, and the position noted at which the star became invisible. Modern photometers, however, are almost always photoelectric.

Photometry. The measurement of the intensity of light. It relates particularly to the magnitudes of stars and remote galaxies.

Photomultiplier. A complicated form of *photoelectric cell, in which the original current produced is magnified many times.

Photon. The smallest 'unit' of light.

Photosphere. The bright surface of the *Sun. Its temperature is 6000 degrees Centigrade.

Piazzi, Giuseppe (1746–1826). Discoverer of the first asteroid, *Ceres, in 1801. He was Director of the Sicilian observatory at Palermo, and compiled an excellent star catalogue.

Pic du Midi Observatory. French observatory in the Pyrénées, at an altitude of 9400 feet. Seeing conditions are excellent, and superb planetary work has been carried out there.

Pickering, E. C. (1846–1919). Pioneer American stellar spectroscopist; one-time Director of the Harvard College Observatory.

Pickering, W. H. (1858–1938). Eminent American lunar and planetary observer. He discovered Saturn's outermost satellite, Phœbe (1898).

Plage, Solar. Large calcium-vapour area surrounding a *sunspot.

Planetarium. An instrument used to show an 'artificial sky' on the inside of a large dome, and to reproduce celestial phenomena of all kinds. The first modern-type projector was installed at Jena, in what is now East Germany, in 1923. This projector was manufactured by the optical firm of Carl Zeiss, and there are other Zeiss planetaria in various cities. Of British planetaria, that at the National Maritime Museum has a Spitz projector, built in America; the planetarium in Baker Street has Zeiss equipment, and the projector of the Armagh Planetarium, in Northern Ireland, is of Japanese manufacture.

Planetaria are becoming extremely popular, and as educational aids in astronomy they are unrivalled. Schools make extensive use of them, particularly in the United States and the Soviet Union.

Planetary Nebula. The name is misleading, since a planetary nebula is neither a planet nor a true nebula! It consists of a faint central star surrounded by an immense gaseous shell, so that it looks like a tiny luminous disk or ring. More than 300 are known in our Galaxy, but even the brightest of them, the Ring Nebula in Lyra (M.57), is much too faint to be seen with the naked eye. Planetary nebulæ are extremely rarefied, and of low mass; they are apparently in a state of expansion. The central stars are always extremely hot, with temperatures of the order of 100,000 degrees Centigrade; since they appear faint they must therefore be small by stellar standards. It is possible that a planetary nebula is produced when a star has entered the giant branch of the *Hertzsprung-Russell Diagram, and has become a Red Giant; it may 'shed' the outer part of its atmosphere, so that the shell of a planetary nebula represents the discarded outer layers of the giant star and the faint central star is nothing more nor less than the core of the old giant.

Planetoid. Alternative name for an asteroid or *minor planet.

Planets. The most important members of the *Solar System (excluding the Sun, of course). They are non-luminous, and move round the *Sun at various distances in various periods. Data concerning the nine known planets are conveniently summarized in a table, on page 125.

At first sight a planet looks like a star, but unlike a star it moves appreciably from night to night, and the ancient Greeks knew quite well that a planet and a star are entirely different. *Mercury, *Venus, *Mars, *Jupiter and *Saturn were known in prehistoric times; of the rest, *Uranus was discovered by Herschel in 1781, *Neptune by Galle in 1846, and *Pluto by Tombaugh in

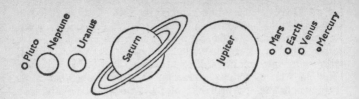

Fig. 60. *Comparative sizes of the planets*

1930 (Fig. 60). (In the last two cases, the main credit was due to the mathematicians, Le Verrier for Neptune and Lowell for Pluto.) There may well be a still more remote planet beyond Pluto, but if so it is bound to be very faint and hard to detect.

It is evident that the planets are divided into two main groups (Fig. 61.) The inner group is made up of relatively small worlds (Mercury to Mars); then comes a wide gap, in which most of the *minor planets move; and beyond this are the four giants (Jupiter to Neptune). Pluto, with its unusual orbit, does not seem to fit into the general plan, and it may be nothing more than an ex-satellite of Neptune. The current ideas about the origin of the planets are discussed briefly under the heading *Solar System.

There is every reason to suppose that other stars, too, have their own planet-families, but no proof is possible by direct observation with our present-day telescopes, since even a large planet moving round a nearby star would be too faint to be seen. However, such a planet might affect its parent star's

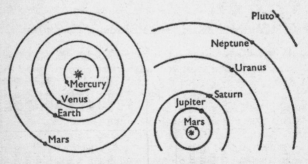

Fig. 61. *Orbits of the planets, to scale. Left, the inner planets. Right, the outer planets*

124

Planet	Mean distance from Sun, in miles	Sidereal Period	Synodic Period (days)	Orbital Eccentricity	Orbital Inclination (deg sec)		Mean Orbital Velocity (miles/sec)	Axial Rotation (days hours)		Inclination of Axis (degrees)
Mercury	36,000,000	88 days	115·9	0·206	7	0	29·8	58	7	0
Venus	67,200,000	224·7 „	583·9	0·007	3	24	21·8	243		178
Earth	92,957,000	365·3 „	—	0·017	0		18·5	23h	56m	23·5
Mars	141,500,000	687 „	779·9	0·093	1	51	15·0	24	37	25·2
Jupiter	483,300,000	11·9 years	398·9	0·048	1	18	8·1	9	51	3·1
Saturn	886,100,000	29·5 „	378·1	0·056	2	29	6·0	10	14	26·7
Uranus	1783,000,000	84·0 „	369·7	0·047	0	46	4·2	10	48	98
Neptune	2793,000,000	164·8 „	367·5	0·009	1	46	3·4	±14		29
Pluto	3666,000,000	248·4 „	366·7	0·246	17	10	2·9	6d	9h	?

Planet	Equatorial Diameter (miles)	Mass (Earth=1)	Volume (Earth=1)	Density (Water=1)	Surface Gravity (Earth=1)	Escape Velocity (mil/sec)	Albedo (per cent)	Oblateness	Maximum surface temperature (°F)	Number of Satellites
Mercury	3000	0·05	0·06	5·5	0·37	2·6	7	0·0	+770	0
Venus	7700	0·81	0·85	5·0	0·89	6·4	76	0·0	+900	0
Earth	7927	1	1	5·5	1	7	36	0·003	+140	1
Mars	4219	0·11	0·15	3·9	0·39	3·2	16	0·005	+80	2
Jupiter	88700	318	1312	1·3	2·64	37·1	73	0·062	−200	13
Saturn	75100	95	763	0·7	1·16	22·0	76	0·096	−240	10
Uranus	29300	15	50	1·7	1·11	13·9	93	0·06	−310	5
Neptune	31200	17	43	1·8	1·21	15·1	84	0·02	−360	2
Pluto	3700?	?	?	?	?	?	?	?	?	0

*proper motion, causing the star to 'wobble' very slightly. In this way several invisible bodies have been found. Barnard's Star, which is a red dwarf only 6 light-years away, has been carefully studied by P. van de Kamp and his colleagues at the Sproule Observatory in America, with interesting results. Van de Kamp considers that there may be two planets in the system of Barnard's Star, each comparable in mass with Jupiter. The fainter component of the *binary system 61 Cygni may also have a planetary companion, and there are a few other 'suspects', including Epsilon Eridani, which is only 10½ light-years away and is not so very unlike our Sun, though somewhat smaller and cooler.

Pleiades. The famous star-cluster in Taurus (the Bull), conspicuous with the naked eye (Fig. 62). It is often known as the Seven Sisters, since under good conditions a normal-sighted person can make out seven individual stars. Many more stars are shown telescopically, and it is thought that the whole cluster contains more than 200 members, together with a beautiful reflection *nebula shown only on long-exposure photographs. The distance of the Pleiades is some 400 light-years.

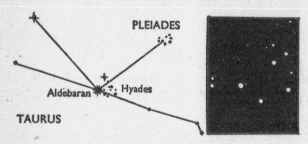

Fig. 62. *The Pleiades*

Several of the stars have been given separate names. The leading member is Alcyone, of the third magnitude; then follow Electra, Atlas, Merope, Maia, Taygete, Celæno, Pleione and Asterope. Pleione is a remarkable *shell-star, of great interest to astrophysicists, though when seen in a telescope it looks ordinary enough.

The chief stars of the Pleiades are hot and white. It has been estimated that they were produced out of interstellar material only about 60,000,000 years ago, so that by cosmical standards they are very young indeed.

Plough. Unofficial name for the main pattern of *Ursa Major (the Great Bear).

Pluto. The outermost of the nine known planets (Fig. 63). It was identified by Clyde Tombaugh, at the Lowell Observatory in Arizona, in March 1930, after its position had been predicted by Percival Lowell, founder of the Observatory. Lowell had given the position because of the *perturbations exerted by Pluto upon the orbits of Neptune and (particularly) Uranus, though in fact Pluto was not located until fourteen years after Lowell's death.

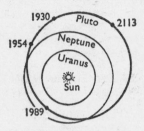

Fig. 63. *Orbit of Pluto*

The diameter of Pluto is probably about 3700 miles, so that it is considerably smaller than the Earth. Unfortunately no exact measures are possible, since even in a powerful telescope Pluto looks like a mere speck of light. The brightness varies regularly over a small range, and from this it has been found that the axial rotation is 6 days 9 hours.

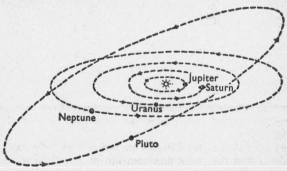

Fig. 64. *Orbits of the outer planets (not to scale)*

127

The orbit is peculiar, since it is tilted or inclined to the main plane of the Solar System at a decided angle (Fig. 64). When near *perihelion, Pluto comes well within the orbit of Neptune, though there is no risk of a collision between the two bodies. There are suggestions that Pluto is an ex-attendant of Neptune, which was broken free for some reason or other and has moved off along an independent orbit, but this is merely a plausible guess, and there is no way of proving whether the idea is right or wrong.

Pluto will reach perihelion in 1989. Already it is considerably brighter than when Tombaugh discovered it in 1930, and a moderate telescope will show it, though it looks exactly like a faint star. At present it lies near the boundary, between Virgo and Coma Berenices; it will remain in this region for some years since its movement against the starry background is very slow.

There seems no chance of life upon Pluto. The cold must be intense, and no atmosphere of any sort has been found as yet.

Poles, Celestial. The north and south points of the *celestial sphere. The north celestial pole is marked approximately by Polaris; there is no prominent south polar star, the nearest naked-eye object to the polar point being the fifth-magnitude Sigma Octantis.

Pond, John (1767–1836). The sixth Astronomer Royal (1811–1835). His early administration was fruitful, and he was an excellent observer; but ill-health later handicapped him, and he was asked to resign office.

Populations, Stellar. Two main types of star regions. Population I consists of brilliant, hot white stars, together with a great deal of interstellar material in the form of dust and gas; the brightest stars of Population II are Red Giants, and there is relatively little interstellar material. Since Red Giants are 'old' stars which have used up their main energy resources, it follows that Population II is older than Population I.

No hard and fast boundaries can be laid down, but the arms of spiral *galaxies are mainly of Population I, while the central parts of spiral galaxies, as well as elliptical galaxies and *globular clusters, are mainly of Population II.

Position Angle. The apparent direction of one object with reference to another, measured from the north point of the main object through east (90 degrees), south (180), and west (270) back to north (000 or 360 degrees). Position angles of double stars are carefully measured. One of these, the famous Mizar in the tail of the Great Bear, is shown in the diagram (Fig. 65); the position angle is 150 degrees.

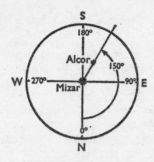

Fig. 65. *Position angle*

Præsepe. Messier 44; naked-eye open cluster in Cancer (the Crab). It is often nicknamed the Beehive. Apart from the *Pleiades, it is possibly the finest of all open clusters.

Precession. The apparent slow movement of the celestial poles.

The Earth is not a perfect globe; its equator bulges slightly, and the Moon pulls upon this bulge. The result is that the Earth's axis seems to describe a small circle, in the manner of a gyroscope that is starting to topple (Fig. 66). The circle on the *celestial sphere is only 47 degrees in diameter, and takes 25,800 years to complete, but the effects are important. Because the celestial poles shift, the celestial *equator also moves. This in turn shifts the position of the vernal *equinox or *First Point of Aries, which is where the equator cuts the *ecliptic. Since the *right ascensions and *declinations of stars are measured from the First Point of Aries and the celestial equator, these, too, will alter slightly from year to year.

The First Point of Aries moves by 50 seconds of arc yearly along the ecliptic, from west to east. Since ancient times, this slow motion has taken the First Point out of Aries into the neighbouring constellation of Pisces (the Fishes), though the name has not been changed. In ancient times, too, the north

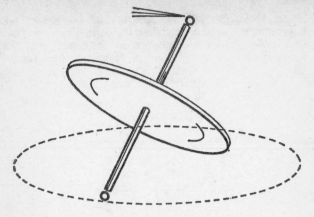

Fig. 66. *Precession*

celestial pole lay close to the star Thuban in Draco (the Dragon). At the moment Polaris occupies the position of honour, but by AD 12,000 we shall have a much more brilliant pole star – Vega in Lyra.

Precession of the equinoxes has been known for a long time. It was discovered by the Greek astronomer *Hipparchus in the second century BC.

Prime Meridian. The meridian on the Earth's surface which passes through the instrument known as the Airy *transit, at old Greenwich Observatory. It is taken as longitude 0 degrees, and marks the boundary between the eastern and the western hemispheres.

The decision to take this particular meridian as 0 degrees was reached by international agreement in the 19th century. The Airy transit instrument is so called because it was erected by Sir George Airy, who was Astronomer Royal for many years.

Prism. A triangular or wedge-shaped block of glass. Light passing through it is split up, because the different wavelengths in it are bent or *refracted* by different amounts (Fig. 67); for instance, blue light is bent more than red. Prisms are essential parts of most instruments based on the principle of the *spectroscope*, though in some cases they may be replaced by *diffraction gratings.

In a *Newtonian reflector, a special prism may replace the

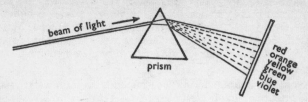

Fig. 67. *Production of a spectrum*

more conventional flat mirror used to send the light from the main speculum into the side of the tube. If this is done, there is of course no splitting-up of the light into a spectrum, since parts of the prism are silvered, though a flat mirror is far better than a prism for this purpose.

Prominences. Masses of glowing gas, chiefly hydrogen, rising from the Sun's surface. Many are associated with sunspots, and they are of great size; the length of an average prominence is over 100,000 miles. They are of two main types. *Eruptive prominences* are in violent motion, while *Quiescent prominences* are relatively calm, so that they may persist for relatively long periods of several weeks.

Prominences can be seen with the naked eye only when the Sun is hidden by the Moon during a total solar *eclipse, but instruments such as the *Lyot filter can show them at any time. They were formerly known as Red Flames, but the name was misleading, and is never used nowadays.

Proper Motion. The individual motion of a star on the *celestial sphere. All stellar proper motions are very slight, because the stars are so remote; the *constellation patterns remain to all intents and purposes unaltered from century to century, though over a very long period the tiny proper-motion shifts will cause the patterns to alter (Fig. 68).

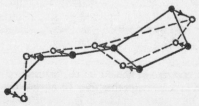

Fig. 68. *Proper motions of stars in the Great Bear*

Barnard's Star, a dim Red Dwarf 6 light-years away, has the largest proper motion known, amounting to one minute of arc every 6 years. It will therefore take 180 years to move by an amount equal to the apparent diameter of the Moon. The proper motions of distant stars are too slight to be measured at all.

Proton. A fundamental particle, with a positive electrical charge. The nucleus of the simplest *atom known (hydrogen) is made up of one proton; the nuclei of other atoms are made up of protons and *neutrons. Since a neutron has no electric charge, it follows that the total charge of an atomic nucleus is positive. This positive charge is balanced out by the combined negative charges of the *electrons moving round the nucleus, so that the complete atom is electrically neutral.

Proton-Proton Reaction. One of the ways in which a star may produce its energy. As with the *carbon-nitrogen cycle, the final result is that hydrogen is changed into helium, with loss of mass and the release of energy. It is now known that the proton-proton reaction plays the main role in stars such as the *Sun.

Proxima. The nearest star beyond the Sun; a faint red dwarf. It is a member of the system of *Alpha Centauri.

Ptolemæus, Claudius (Ptolemy) (circa AD 120–180). Last great astronomer of Classical times. Much of our knowledge of ancient astronomy comes from his book, which has come down to us in its Arab translation as the *Almagest*.

Ptolemaic System. The old plan of the Universe, according to which the Earth lay motionless in the centre with all other bodies moving around it (Fig. 69). It was described in detail by Ptolemy, around AD 150. Not until more than a thousand years later was it first challenged and then replaced by the *heliocentric theory, which was revived by Copernicus in 1543 and which placed the Sun, not the Earth, in the central position.

Pulkova Observatory. Major Russian observatory, near Leningrad. It includes a 30-inch refractor. It was destroyed during the war, but has now been rebuilt, though it must be admitted that observing conditions there are not good.

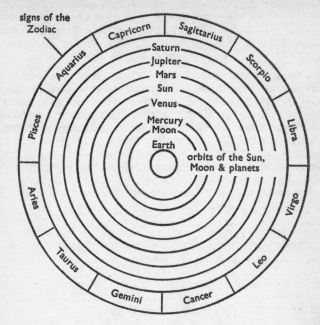

Fig. 69. *The Ptolemaic System*

Pulsar. A rapidly-varying radio source, now believed to be a *neutron star.

The first pulsar was discovered in 1967, by Miss Jocelyn Bell, a member of the research team led by Professor Anthony Hewish at Cambridge. The source fluctuated so rapidly, and with such absolute regularity, that for a brief period there were serious suggestions that it might be artificial. This intriguing idea was soon abandoned, and by now dozens of pulsars are known. The most famous is that in the *Crab Nebula; it has a period of 0·03 seconds, and has been identified with a faint optical object flashing in exactly the same period, so that coincidence can be ruled out. The Crab is, of course, a supernova remnant, and probably the same is true of the other pulsars.

Pulsars are contained in our Galaxy (the Crab lies at a distance of 6000 light-years). Despite the similarity of name, they should not be confused with quasars, which are well outside our Galaxy and are immensely powerful.

133

Q

Quadrant. An astronomical measuring instrument used in former times. It consisted of an arc graduated into 90 degrees, together with a sighting pointer; it was used to measure the apparent positions of stars and other celestial bodies. Since quadrants were often large, they were attached to walls, and were then known as *Mural Quadrants* (Latin *murus*, a wall).

Quadrantids. Short, sometimes spectacular *meteor shower of early January. It is named because the radiant lies in the now-rejected constellation of Quadrans (the Quadrant), not included in Boötes.

Quadrature. When the Moon or a planet is at right angles to the Sun as seen from Earth, it is said to be in quadrature. Thus the Moon is in quadrature when at half-phase.

Quantum. The energy possessed by one *photon of light. A quantum is therefore the smallest amount of light-energy which can be transmitted at any given wavelength.

Quasar. A very remote, incredibly luminous object. The name 'quasar' has now generally superseded the original 'quasi-stellar object' or QSO.

In 1963 Maarten Schmidt, at the Palomar Observatory in America, examined the spectrum of an object which seemed to coincide in position with a strong radio source. The object looked like a star when seen through a telescope, but the spectrum showed it to be something quite different. There was no similarity to the spectrum of a normal star, and there was a tremendous *red shift, which presumably meant that the object was receding at high velocity. This, in turn, meant that it must be remarkably luminous; some two hundred times brighter than a whole galaxy!

The first quasar caused intense interest among astronomers, and a search was made for other objects of the same sort. A number of quasars soon became known, all with luminosities of the same order. One particular quasar was found to be receding at some 150,000 miles per second, which would place its distance from us at at least 6000 million light-years. Matters were further complicated in 1965, when A. Sandage at Palomar reported 'quiet quasars' which were just as powerful, but did not emit detectable radio waves. However, the nature of these objects is still uncertain.

Quasars (including quasi-stellar galaxies) have very small angular diameters, and the problem at present is to decide how a relatively small body, only a few light-years in diameter, can radiate as fiercely as a million million Suns put together. Various theories have been put forward, but none seems convincing, and all we can say is that the quasars must be drawing upon some source of energy about which we know nothing at all – unless, of course, they are much closer to us than the red shifts indicate, which is not impossible. Incidentally, some quasars vary in light over comparatively short periods.

Quasars had been photographed often enough during the past fifty years or so, but not until 1963 was it realized that they were anything but faint blue stars. Their identification is one of the most important astronomical developments of the 20th century, but as yet they remain highly enigmatic.

R

Radar Astronomy. A new branch of astronomical science, dating only from the end of the war. It depends on the principle of radar; a pulse of energy is transmitted, and 'bounced back' from a distant object, so that the 'echo' is received and measured. The only everyday comparison is to picture what happens when a tennis-ball is thrown against a wall, and caught on the rebound. This analogy is by no means accurate, but it does serve to give the general idea.

Radar echoes have been obtained from various bodies in the Solar System, including the Moon, Venus, Mars, Mercury and the Sun. The most important result has been obtained with regard to Venus. Since a radar pulse moves at the same velocity as light (186,000 miles per second), the time-delay between the transmission of the pulse and the reception of the echo from Venus gives the distance which the pulse has travelled – and hence the distance of Venus itself may be found. Once this is known, *Kepler's Laws can be used to give a precise value for the *astronomical unit or Earth-Sun distance.

Meteors can also be studied by radar, since for this purpose a meteor trail behaves in the same way that a solid body would do. By now, radar observation of meteors has largely replaced the older visual methods.

Radial Velocity. The towards-or-away movement of a celestial body, measured by the *Doppler effect shown in its spectrum. Radial velocity is said to be positive if the object is receding, negative if the object is approaching. All galaxies except those of the *Local Group have large positive radial velocities.

Radiant. The point in the sky from which the *meteors of any particular shower seem to radiate. For instance, the spectacular August shower has its radiant in the constellation Perseus, which is why the meteors are known as the Perseids.

Radio Astronomy. A branch of astronomy which began less than forty years ago, but which has now become of the utmost importance.

In 1931 K. Jansky, an American radio engineer, was using a special form of aerial to study static electricity when he found that he was picking up radio waves from the sky. They proved to come from the region of the *Milky Way in the constellation Sagittarius (the Archer), now known to indicate the direction of the centre of the *Galaxy. Jansky's work was followed up by an American amateur, Grote Reber, who built the first real *radio telescope in 1939, but only after the war did radio astronomy become a major study.

Radio waves are electromagnetic vibrations, and are of the same basic nature as visible light, but they do not affect our eyes, and have to be collected and studied by instruments which are really in the nature of large aerials. After 1947, many radio sources in the sky were found. Surprisingly enough, the individual stars did not seem to be radio emitters, and the old term of 'radio star' was soon dropped.

Some sources, such as the *Crab Nebula, proved to be the remnants of old *supernovæ, and the Sun was also found to be a radio source; later, emissions were detected from the planet Jupiter. Yet most of the sources were much more remote, lying well beyond our own Galaxy. There are some external galaxies which are extremely powerful at radio wavelengths, though as yet we do not pretend to know why.

All sorts of developments have been made possible by the new methods. Without radio astronomy nothing would be known, for instance, of the *quasars. We have also cleared up one vital point about our own *Galaxy. Between the stars there are clouds of cold hydrogen, quite invisible optically, but which emit radio waves at 21·1 centimetres; the distribution of the clouds has been studied, and it has been found that the *Galaxy is spiral in form.

It is also worth noting that radio waves can be received across immense distances, so that in this way we can reach further into space than is possible with ordinary telescopes. Many of the radio sources now being studied are so remote that not even the world's largest optical telescopes will show them at all.

Radio Telescope. An instrument used for studying the radio waves from space. The name is not an apt one, since a radio telescope does not produce a visible picture, and one cannot look through it, as so many non-scientists fondly imagine! The radio waves are collected and focused, and the result appears as a trace on a graph.

It is also possible to convert the energy into sound – hence the hackneyed term *radio noise* – but the actual sound is produced in the equipment, and does not come direct from space. Where there is no air to carry sound-waves, there can be no noise; and the Earth's atmosphere extends upward for only a few hundreds of miles.

Some radio telescopes are *paraboloids*, and take the form of huge metal dishes. The best-known fully steerable paraboloid is the 250-foot dish at Jodrell Bank, near Manchester, which was completed in 1955. Other instruments, such as those at Cambridge, depend upon the principle of interference, and are known as *radio interferometers*. In fact, radio telescopes take many forms, each of which is suitable for some particular branch of research.

Radio Star. When radio astronomy began, it was thought – rather naturally – that distance sources would be identified with stars. However, it was soon found that the sources were not stars; most of them were either the remnants of *supernovæ, or else external galaxies. The name *radio star* has therefore been dropped.

Apart from the Sun, the only true stars from which radio emission has been detected are some faint red dwarfs of the *flare star type and a few more distant stars, such as the bluish companion of Antares. No doubt all stars do in fact emit at radio wavelengths – it would be absurd to suppose that our Sun is exceptional – but the emissions are too weak to be detected with our present equipment. Were the Sun removed to a distance of a few light-years, we would not be able to record its radio waves even with the Jodrell Bank 'dish'.

R Coronæ Borealis Stars. Rare, remarkable variable stars, which remain normally at maximum but which undergo sudden, unpredictable falls to minimum. R Coronæ is the brightest member of the class.

Red Shift. When a luminous body is receding, the *Doppler Effect means that its spectral lines will be shifted over to the red or long-wave end of the spectrum. The greater the shift, the greater the speed at which the body is moving away. Apart from the members of the *Local Group, all external galaxies show red shifts, which is taken to mean that the whole universe is in a state of expansion.

Reflector. An optical telescope in which the light from the object under study is collected by a curved mirror. Details are given under the headings *Cassegrain reflector, *Gregorian reflector,

*Herschelian reflector, *Newtonian reflector, *Maksutov telescope and *Schmidt telescope.

Refraction. The 'bending' or change of direction of a ray of light when passing through a transparent substance. For instance, light is refracted when it passes through the *object-glass of a telescope. The Earth's atmosphere also causes refraction; an object close to the horizon will be seen as higher up than it really is, and the effect may amount to more than half a degree, so that the Sun or full Moon may sometimes be visible when it is theoretically below the horizon. When on the point of setting, the lower limb of the Sun will be more affected than the upper limb, so that the Sun's disk will appear obviously flattened (Fig. 70).

Fig. 70. *Apparent distortion of the low Sun*

Refractor. A telescope which collects its light by means of a lens. The light passes through the main lens or *object-glass, and is brought to focus, where the image is magnified by an *eyepiece (Fig. 71).

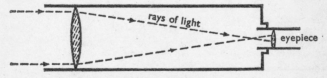

Fig. 71. *Principle of the refractor*

Astronomical refractors have compound object-glasses, as otherwise an object such as a star would be marred by false colour. (In fact, the false colour nuisance can never be completely cured.) A refractor is more effective than a *reflector of the same aperture, and is convenient to use, but is much more expensive.

The first astronomical refractors were made during the 17th century. Pride of place today goes to the 40-inch at Yerkes Observatory, in America; it is unlikely that a larger refractor will be built, partly because it is much easier to make a reflector with superior light-grasp, and partly because a giant lens tends to distort under its own weight, making it useless. This trouble does not apply to a reflector, since a mirror can be supported at its back, whereas the object-glass of a refractor has to be supported round its edge.

Recurrent Nova. A star which has been known to show more than one nova-like outburst. The best-known example is T Coronæ, in the constellation of Corona Borealis (the Northern Crown). Normally it is very faint, but twice, in 1866 and again in 1946, it has flared up to naked-eye visibility for a while.

Regression of the Nodes. The *nodes of the Moon's orbit move slowly westward, making one full circuit of the orbit in 186 years. This *regression* is due to the gravitational pull of the Sun.

Resolving Power. The ability of a telescope to separate objects which are close together. The larger the telescope, the better its resolving power. For instance, with a 1-inch telescope the star Castor, in Gemini (the Twins) appears as a single mass; with a 6-inch telescope, two components are seen separately.

Radio telescopes have very poor resolving power compared with optical telescopes, and it is never easy to tell whether a radio source is single or whether it is made up of several parts.

Retardation. The difference in the time of moonrise between one night and the next. It may exceed one hour, but at the time of *Harvest Moon it may be reduced to as little as a quarter of an hour.

Retrograde Motion. In the Solar System, a body which moves round the Sun in a direction opposite to that of the Earth is said to have retrograde motion. Many comets behave in this way (including the famous Halley's Comet), but no retrograde planet or minor planet is known. The term is also applied to the satellites of the planets; for instance, nine of Saturn's satellites have direct motion, while the other (Phœbe) is retrograde.

Phœbe may thus be compared with a car which is moving the wrong way round a roundabout.

Another meaning of the term refers to the apparent movements of the planets. Usually, a planet moves eastward against the starry background, but we are observing from the Earth, which is itself in motion, so that the planets seem to move in a westward or retrograde direction at various times.

Reversing Layer. The gaseous layer above the bright surface or *photosphere of the Sun. Shining on its own, the gas would produce bright spectral lines; but as the photosphere makes up the background, the lines are reversed, and appear as dark. (Further details are given under the heading *Spectroscope.) Strictly speaking, the whole of the Sun's *chromosphere is a reversing layer.

Rhea. The fifth *satellite of Saturn. It is bright enough to be seen with a small telescope.

Rigel. The brightest *star in *Orion. It has a B-type spectrum, and is very luminous and remote, with an estimated luminosity 50,000 times that of the Sun.

Right Ascension. The angular distance of a star from the *First Point of Aries or vernal *equinox, measured westward (Fig. 72). It is usually given in hours, minutes and seconds of time. The First Point of Aries must reach its highest point (culmination)

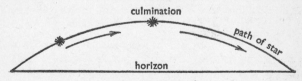

Fig. 72. *Culmination*

once a day; the right ascension of a star, or other celestial body, is the time-difference between the culmination of the First Point of Aries, and the culmination of the star.

For instance, the bright star Aldebaran in Taurus (the Bull) culminates 4h 33m after the First Point has done so. The right ascension of Aldebaran is therefore 4h 33m.

The right ascensions of bodies in the Solar System change

141

quite rapidly. The right ascensions of stars are, however, constant – apart from the small, slow changes due to *precession.

Rill (German, *rille*). Crack-like feature on the surface of the Moon. Rills are alternatively known as clefts.

Ring Micrometer. A form of measuring instrument, in which no moving wires are used as in the *filar micrometer. The ring micrometer is the simpler of the two, but is not so accurate.

Ring Nebula. Messier 57, the famous planetary nebula in Lyra.

Roche Limit. The distance from the centre of a planet within which a second body would be broken up by the planet's gravitational pull. This applies only to an orbiting body which has no appreciable 'gravitational cohesion', so that strong, solid objects such as artificial satellites may move safely well within the Roche limit for the Earth. The Roche limit lies at 2·44 times the radius of the planet from the centre of the globe, so that for Earth the limit is about 5700 miles above ground level.

All known planetary satellites lie outside the Roche limits of their primaries. Saturn's ring-system, however, lies inside the Roche limit for *Saturn, and it has been suggested – though without proof – that the rings are nothing more nor less than the fragments of a former satellite which was broken up when it moved inside the Roche limit.

Rocket Astronomy. One of the most spectacular branches of modern science, in which rockets are used to explore the nearer parts of the Solar System.

A rocket functions by what is termed the *principle of reaction*. Its propellants produce hot gases, which are sent out of the exhaust of the vehicle, so propelling the vehicle itself in the opposite direction; the rocket 'pushes against itself', so to speak, and works well when there is no atmosphere surrounding it. The basic principle is just the same as that of an ordinary November the Fifth rocket.

More than sixty years ago the Russian pioneer, K. E. Tsiolkovskii, realized that a rocket will work in vacuum. In 1926 an American, R. H. Goddard, fired the first rocket powered by liquid propellants, which are much more efficient than solid fuels of the gunpowder type, and are also much easier to control. During the war, the Germans developed rockets for military purposes; and after the end of hostilities, rocket research was continued both for scientific use and for military misuse.

Early rockets carried instruments to study the upper layers of the Earth's atmosphere. In 1955 the Americans announced that they proposed to launch an *artificial satellite by means of rocket power, and in 1957 the Russians actually did so. Since then, many satellites have been sent up, and have provided an immense amount of information which could not have been gained in any other way. The first man to make a true space-flight was Yuri Gagarin, of the Soviet Air Force, in 1961.

During the 1960s rockets became steadily more efficient, and were used to take complicated equipment above the Earth's atmosphere; for instance, *X-ray astronomy depends entirely upon rocketry, since the X-rays from space are blocked by the atmosphere of the Earth, and cannot penetrate to ground level.

Manned probes have been sent to the Moon and unmanned probes have gone beyond Venus and Mars. A manned 'research laboratory' has been maintained in orbit round the Earth. There is no doubt that rocket astronomy will remain an essential part of modern research.

Rømer, Ole (1644–1710). Danish astronomer; he was the first to measure the velocity of light (in 1675) and also invented various astronomical instruments, including the *transit instrument.

Rosse, 3rd Earl of (1800–1867). Irish amateur astronomer, who built what was then the world's largest telescope – the 72-inch reflector – and used it to discover the spiral nature of the galaxies. The telescope tube can still be seen at Birr Castle.

Rosse, 4th Earl of (1840–1908). He continued his father's work at Birr, and was the first to make an accurate measurement of the tiny quantity of heat sent to us by the Moon.

Royal Observatory, Edinburgh. The senior Scottish observatory, at Blackford Hill, Edinburgh.

RR Lyræ Variables. Regular *variable stars whose periods are very short, between 1¼ hours and 30 hours. All seem to have about the same real luminosity, not much less than 100 times that of the Sun, and so they can be used for distance-gauging in the same way as the *Cepheids. Many of them are found in star-clusters, and they were once called cluster-Cepheids, but they are now known as RR Lyræ variables, since the star RR

Lyræ is the best-known member of the class. No RR Lyr variable appears bright enough to be seen with the naked eye.

Russell Diagram. See *Hertzsprung-Russell Diagram.

Russell, Henry Norris (1877–1957). Leading American astrophysicist; he is remembered chiefly for the *Hertzsprung-Russell Diagram.

RW Aurigæ Variables. Irregular variable stars; their changes in light are rapid and unpredictable.

S

Saros. A period of 18 years 11·3 days, after which the *Earth, *Moon and *Sun return to almost the same relative positions. It is due to the *regression of the nodes of the Moon's orbit. The Saros can be used to predict eclipses; it is usual for an eclipse to be followed by a similar eclipse 18 years 11·3 days later, though the slight differences from one Saros to another mean that the eclipses are not identical. For instance, the total solar eclipse of 1927 was total in parts of England, but the 'return' eclipse of 1945 was not.

Satellites. Minor bodies which move round some of the planets in the Solar System (Fig. 73). The Moon is usually taken to be a satellite of the Earth, though, as noted under the heading *Moon, it is better to regard the Earth-Moon system as a double planet. *Mars has two satellites, *Jupiter fourteen, *Saturn ten, *Uranus five and *Neptune two, while *Mercury, *Venus and *Pluto do not seem to have any. Data are best given in a table, on page 146.

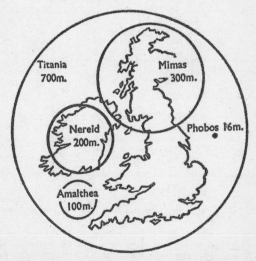

Fig. 73. *Sizes of some satellites, compared with Britain*

Satellite	Mean distance from centre of primary (miles)	Sidereal Period (d. h. m.)	Eccentricity of Orbit	Inclination of Orbit	Diameter (miles)	Maximum Magnitude

SATELLITE OF EARTH

Moon	238,857	27 7 43	0·055	5°9′	2160	−12·5

SATELLITES OF MARS

Phobos	5,800	7 39	0·017	2°	16	10
Deimos	14,600	1 6 18	0·003	2°	6	11

SATELLITES OF JUPITER

V. Amalthea	113,000	11 57	0·003	0°24′	150	13
I. Io	262,000	1 18 28	0·0	0°	2310	5·5
II. Europa	417,000	3 13 14	0·0	0°	1950	5·6
III. Ganymede	666,000	7 3 43	0·0	0°	3270	5·1
IV. Callisto	1,170,000	16 16 32	0·0	0°	2900	6·3
VI. Himalia	7,120,000	250 16	0·158	27°38′	100	13·7
VII. Elara	7,290,000	259 16	0·207	24°6′	35	17
X. Lysithea	7,300,000	260 12	0·130	29°1′	15	18·8
XIII. Leda	7,700,000	210 14	0·244	28°42′		
XII. Ananke	13,000,000	§625	0·169	147°	14	18·9
XI. Pasiphaë	14,000,000	§700	0·207	164°	19	18·4
VIII. Carme	14,600,000	§739	0·378	145°	35	16
IX.Sinope	14,700,000	§758	0·275	153°	17	18·6
XIV						

SATELLITES OF SATURN

Janus	98,000	17 58	0·0	0°	150?	14·0
Mimas	115,000	22 37	0·020	1°31′	300	12·1
Enceladus	148,000	1 8 53	0·004	0°1′	400	11·6
Tethys	183,000	1 21 18	0·0	1°6′	700	10·6
Dione	234,000	2 17 41	0·002	0°1′	800	10·7
Rhea	327,000	4 12 25	0·001	0°21′	900	9·7
Titan	758,000	15 22 41	0·029	0°20′	3500	8·2
Hyperion	919,000	21 6 38	0·104	0°26′	200	13·0
Iapetus	2,210,000	79 7 55	0·028	14°43′	1800?	9
Phœbe	8,040,000	§550 9	0·163	150°	150	14

SATELLITES OF URANUS

Miranda	80,700	1 19 56	0·01	0°	200	17
Ariel	119.000	2 12 29	0·003	0°	1500	14
Umbriel	166.000	4 3 28	0·004	0°	800	14·7
Titania	272,000	8 16 56	0·002	0°	1500	14
Oberon	364,000	13 11 7	0·001	0°	1500	14

SATELLITES OF NEPTUNE

Triton	219 000	§5 21 3 0		159°57′	3000?	13
Nereid	3,450,000	359	0·76	27°27′	200	19·5

Satellites with retrograde motion are marked §. The five satellites of Uranus (Fig. 82) are also technically retrograde, since they move in ‘he plane of the planet's highly-inclined equator, but are not generally reckoned as such. The diameters and magnitudes of the satellites of Saturn, Uranus

146

and Neptune, and the smaller satellites of Jupiter, are very uncertain; different authorities give different values. According to another set of estimates, the diameters in miles are as follows: Io 2300, Europa 1800, Ganymede 3200, Callisto 3220, Titan 3000, Triton 2300, Iapetus only 690.

The thirteenth and fourteenth satellites discovered in 1974 and 1975 are the latest known additions to the list of satellites in the Solar System.

There may be a basic difference between large satellites such as Titan, Triton and the four main attendants of Jupiter, and the much smaller bodies such as Phobos, Deimos and the outer satellites of Jupiter. The small satellites may be nothing more than captured *minor planets, which would at least account for the *retrograde movement of some of them – notably Phœbe. Nereid, in Neptune's system, has a remarkably eccentric orbit more like that of a comet than a satellite.

Another satellite of Saturn was reported in 1904, and was named Themis; but it has not been seen since, and probably does not exist.

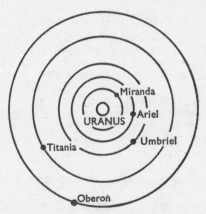

Fig. 74. *Orbits of the satellites of Uranus*

Satellites, Artificial. See *Artificial Satellites.

Saturn. The sixth planet in order of distance from the Sun. Details of its orbit and globe are given under the heading *Planets.

Apart from Jupiter, Saturn is much the largest of the planets; it has an equatorial diameter of over 75,000 miles (Fig. 75), though its quick rotation has made it obviously flattened. It must be made up in the same way as Jupiter, but its density is strangely low, and Saturn 'weighs' less than an equal volume of water would do.

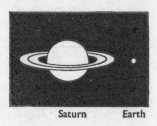

Saturn   Earth

Fig. 75. *Comparative sizes of Saturn and Earth*

When seen through a telescope, the globe of Saturn is yellowish, crossed by belts. There are occasional spots, of which the most famous example of the present century was the white spot discovered by W. T. Hay in 1933, but on the whole the disk seems much less active than that of Jupiter. Moreover, Saturn is not so large, and is much further away from us, so that its details are more difficult to study.

To make up for this, there is a magnificent system of rings, unique in our experience. These rings are made up of large numbers of small particles, moving round Saturn in the manner of dwarf moons; since they lie within the planet's *Roche limit, it has been suggested that they are the fragments of a former satellite which went too close to Saturn and was broken up.

When best placed, the ring-system (Fig. 76) is highly spectacular. At other times, as in 1966, the rings are edge-on to us, and almost disappear, since although they are of great extent they are also very thin; their thickness can hardly be more than ten miles or so.

There are three main rings (Fig. 77), of which two (A and B) are bright, while the third (C) is dusky. Rings A and B are separated by the *Cassini Division*, named in honour of J. D. Cassini, an Italian astronomer who discovered it in 1666.

The dimensions of the ring-system are are follows:

Fig. 76. *Changing aspects of Saturn's Rings*

Width of Ring A	10,000 miles
Width of Cassini Division	1,700 miles
Width of Ring B	16,000 miles
Width of Ring C	10,000 miles
Distance between Ring C and Saturn	9,000 miles

From this it is clear that the entire extent of the ring-system, from one side to the other, is almost 170,000 miles.

Ring B is appreciably brighter than A. Ring C, or the Crêpe Ring, is partially transparent; it was discovered only in 1848, whereas the two main rings had been found in the previous century. (Even Galileo, with his first feeble telescopes, could see them – even though he did not know what they were.) The Cassini Division is visible with a small telescope when conditions are suitable; it is a true gap, which is kept 'swept clear' of ring-particles by the gravitational effects of Saturn's *satellites, particularly Mimas. Another division, *Encke's Division*, has often been recorded in Ring A, but may not be a genuinely clear gap.

A dusky ring outside Ring A has been suspected at various times since 1909, but has never been confirmed, and its existence is doubtful. I have made searches for it, using very large telescopes, but without success. Another dusky ring has been reported inside Ring C, but this too is doubtful. We may know more when the first Saturn probes encounter the planet in 1979.

Of Saturn's ten satellites, one – *Titan – is a large body, and has an atmosphere made up of methane.

Schiaparelli, Giovanni V. (1835–1910). Great Italian planetary observer; he also made notable contributions to cometary studies.

Schmidt, Bernhard (1879–1935). Estonian inventor of the *Schmidt telescope.

Schmidt, Julius (1825–1884). German astronomer, who spent much of his career in Greece as Director of the Athens Observatory. He was mainly concerned with the Moon, and drew a superb lunar map.

Schmidt Telescope (or Schmidt Camera). A type of telescope invented in 1930 by Bernhard Schmidt. It uses a special glass correcting plate, near the top of the tube, as well as a mirror; the mirror itself is spherical. With a Schmidt, relatively wide

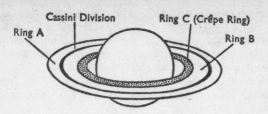

Fig. 77. *Saturn's Ring-System*

areas of the sky may be photographed with one exposure; the definition remains good right up to the edge of the field. Instruments of this kind are not used visually, but have great light-gathering power, so that faint objects such as remote galaxies may be photographed with comparatively short exposures. The world's largest Schmidt is the 48-inch at Palomar, in California.

Schröter Effect. When the planet *Venus is at half-phase, it is said to be at **dichotomy*. Oddly enough, theoretical predictions do not usually agree with observation; when an evening star, and therefore waning, Venus reaches dichotomy earlier than it should do, while when it is waxing in the morning sky dichotomy is late. This effect was first noticed by J. H. *Schröter, a century and a half ago. Presumably it is due to effects of Venus's atmosphere. So far as is known, there is no similar effect for *Mercury, which is of course almost without an atmosphere.

Schröter, Johann Hieronymus (1745–1816). German amateur astronomer; the real founder of *selenography. His observatory, at Lilienthal (near Bremen) was destroyed by invading French troops in 1813.

Schwabe, Heinrich (1789–1875). German amateur observer, who discovered the eleven-year *sunspot cycle.

Schwarzschild, Karl (1873–1916). German astronomer who made fundamental advances in studies of stellar distribution and evolution.

Schwarzschild Radius. The radius that a body must have if its escape velocity is to be equal to the velocity of light.

Consider the Sun, whose present radius is approximately 432,000 miles. If the radius were reduced, without altering the mass, light would find it more and more difficult to escape; and if the solar radius could be reduced to just below two miles, light would be unable to get away at all. We would, in fact, have a *black hole. This is, therefore, the Schwarzschild radius of the Sun. It is named for the German physicist Karl Schwarzschild, who first discussed the concept in 1916. With a body more massive than the Sun, the Schwarzschild radius would naturally be greater.

Scintillation. The official term for 'twinkling'. It is due entirely to the Earth's *atmosphere, and has nothing to do with the stars themselves; a star will twinkle much more strongly when low down than when high up, and may also flash various colours. (This is particularly noticeable with Sirius, the brightest star in the sky.) Planets twinkle less than stars, because they appear as small disks instead of mere points, but *Mercury, in particular, may twinkle quite obviously at times.

Seasons. The Earth's axis is tilted at 23½ degrees to the perpendicular (or, to put matters in another way, the equator is tilted by 23½ degrees to the plane of the orbit). When the northern hemisphere of the Earth is tilted toward the Sun (Fig. 78), as

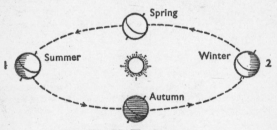

Fig. 78. *The Seasons*

in position 1, it is summer in the northern part of the Earth and winter in the southern; in position 2 in the diagram, conditions are reversed. The fact that the Earth is not always at quite the same distance from the Sun does not have much effect upon the seasons, and in fact the Earth is at its least distance, or *perihelion, in the middle of a British winter!

Secchi, Angelo (1818–1878). Pioneer Italian stellar spectroscopist.

Second of Arc. One-sixtieth of a *minute of arc.

Secular Acceleration. Because of friction produced by the *tides, the Earth's rotation is gradually slowing down – in other words, the day is becoming longer. The average increase of the day amounts to only 0·00000002 seconds, but over a sufficiently long period of time the effect mounts up. Another result of these tidal effects is that the Moon is receding from the Earth at about 4 inches per month.

As each day is 0·00000002 second longer than the previous day, then a century (36,525 days) ago the length of the day was shorter by 0·00073 seconds. Taking an average between then and now, the length of the day was half of this value, or 0·00036 seconds, shorter than at present. But since 36,525 days have passed by, the total error is $36,525 \times 0·00036 = 13$ seconds. Therefore, the position of the Moon, when 'calculated back', will be in error; it will have seemed to have moved too far, i.e. too fast. This is the lunar *secular acceleration.* It shows up during calculations of *eclipses which took place in the distant past.

Seeing. The quality of the steadiness and clarity of a star's image. On a calm night, when there is little twinkling and the star does not seem to be moving about, the seeing is good; telescopically, the disk of a planet will be sharp and steady. Seeing is often bad on a very transparent, brilliantly starlight night, when the images will seem to dance about due to un-steadiness in the Earth's atmosphere.

Selenography. The study of the Moon's surface.

Sextant. An instrument used for measuring the altitude of a celestial body above the horizon.

Seyfert Galaxies. Galaxies with relatively small, bright nuclei. The spectra of the nuclei show emission lines, which must be due to high-temperature gases there; there is evidence of high tur-bulence velocities of several thousands of miles per second, apparently resulting from some violent disturbance. Many Seyfert galaxies are radio sources, and some (such as NGC 4151) are also X-ray emitters. About 1% of all known galaxies are Seyferts. They are decidedly puzzling objects; it has been suggested that they represent the start of an explosive period in the history of a galaxy, or even that Seyfert galaxies and *quasars are different stages of evolution of the same class of object. The original research was carried out by Carl Seyfert during the second world war.

One famous Seyfert galaxy is M.77, in Cetus, which is a radio source, and has a mass estimated at 800,000 million times that of the Sun. It is 52,000,000 light-years from us, and is the most massive galaxy contained in *Messier's catalogue.

Shadow Bands. Wavy lines seen across the Earth just before or just after totality in a total solar *eclipse. They are due to the effects of the Earth's atmosphere. Shadow bands are not seen at every total eclipse, and they are difficult to photograph well.

Shell Stars. Certain very hot white stars which are known to be surrounded by shells of tenuous gas, quite invisible in ordinary telescopes, but detectable because of their effects upon the star's spectrum. Pleione, in the Pleiades, is a good example of a shell star.

Shooting Stars. The common name for *Meteors.

Sidereal Clock. A clock designed so as to keep *sidereal time.

Sidereal Day. This is described under the heading *Day.

Sidereal Month. This is described under the heading *Month.

Sidereal Period (or Periodic Time). The time taken for a planet or other body to make one journey round the Sun (365·2 days in the case of the Earth). The term is also used for a satellite in orbit round a planet.

Sidereal Time. The local time reckoned according to the apparent rotation of the *celestial sphere. The sidereal time is 0 hours when the *First Point of Aries crosses the observer's *meridian – that is to say, when the First Point of Aries culminates, and lies due south. (This applies, of course, to the northern hemisphere of the Earth.) Since the sidereal day is slightly shorter than the solar day, the local sidereal time will not usually be the same as the ordinary civil time. The 24-hour system is used.

The sidereal time for any observer is equal to the *right ascension of an object which lies on the meridian at that time. Thus when Aldebaran (right ascension 4 hours 33 minutes) is on the meridian as seen from Armagh, then the sidereal time at Armagh is 4 hours 33 minutes.

Sidereal Year. This is described under the heading *Year.

Siderite. An iron *Meteorite.

Siderolite. A stony-iron *Meteorite.

Sinope. The ninth satellite of Jupiter.

Sirius. The brightest *star in the sky: Alpha Canis Majoris. It is 26 times as luminous as the Sun, and has a faint *White Dwarf companion.

Skylab. The first true American space-station, launched from Cape Canaveral in May 1973. Three crews manned it in succession, carrying out pioneer scientific work in many fields. The last Skylab astronauts – Carr, Gibson and Pogue – landed back on Earth on 1974 February 8, after having spent a record 84 days on the station. The Skylab experiment proved that men can stay in space for long periods without suffering ill-effects, and it paved the way for all subsequent manned exploration.

Slipher, Earl C. (1883–1964). American astronomer at the *Lowell Observatory. His planetary photographs have seldom been equalled.

Slipher, Vesto M. (1875–1966). Leading American stellar spectroscopist, who succeeded *Lowell as Director of the *Lowell Observatory.

Solar Constant. The unit for measuring the amount of energy received on the Earth's surface, due to solar radiation. It amounts to 1·94 calories per minute per square centimetre. (A calorie is the amount of heat needed to raise the temperature of 1 gramme of water by 1 degree Centigrade.)

Solar Parallax. The trigonometrical *parallax of the Sun. It has a value of 8·79 seconds of arc, giving a mean distance for the Sun of 92,957,209 miles.

Solar System. The system made up of the *Sun, the planets, satellites, comets, minor planets, meteoric bodies, and interplanetary dust and gas. It is a very small part of the universe, and seems important to us only because we happen to live inside it!

The origin of the Solar System is not definitely known; all we can say is that the age must be at least 5000 million years. For a long time Laplace's *Nebular Hypothesis was accepted, but when fatal objections to it were raised an alternative idea was put forward by Chamberlin and Moulton, of the United States. This involved a close approach of a wandering star, which tore material out of the Sun; when the passing star moved off, the torn-away material condensed into planets. The theory was developed by Sir James Jeans, and became popular for a while, but it too had so many weak links that it had to be given up. Today there are various ideas about how the planets were formed. The most plausible is due to the German physicist C. von Weizsäcker, who believes that the Sun and planets were built up by accretion from material making up a 'solar nebula'. At all events, there is no reason to suppose that planetary systems are rare; what can happen to the Sun can also happen to other stars. Unfortunately, proof will be hard to obtain.

Solar Time, Apparent. The local time, reckoned according to the *Sun. Noon occurs when the Sun crosses the observer's *meridian, and is therefore at its highest in the sky.

Solar Tower. A tower with a *cœlostat placed at the top. The cœlostat reflects the Sun's light vertically downward, where it may be studied with spectrographic equipment. The solar tower is a convenient arrangement, since the main equipment can remain fixed in position; the only movable part is the cœlostat.

Solar Wind. A steady flow of atomic particles, streaming out from the *Sun in all directions. The solar wind was detected by space-probes, and most vehicles sent out toward the planets carry instruments to study it.

Solstices. The times when the Sun is at its northernmost point in the sky, declination $23\frac{1}{2}°$N., around June 22 (Summer Solstice; midsummer in the northern hemisphere) and at its southernmost point, declination $23\frac{1}{2}°$S., around December 22 (Winter Solstice; midwinter in the northern hemisphere). The actual dates of the solstices vary somewhat, because of the calendar irregularity due to Leap Years.

Solstitial Colure. The *hour circle passing through the *solstices. It is described under the heading *Colures.

Solstitial Points. The points along the ecliptic when the Sun reaches its maximum declination north or south ($23\frac{1}{2}°$ in each case). Obviously, the solstitial points are 90 degrees away from the *equinoxes.

Space Research. An alternative name for the relatively new science of *astronautics.

Specific Gravity. The density of any substance, compared with that of an equal volume of water. For instance, the specific gravity of the Earth is 5·5, taking water=1. The specific gravity of the Sun is only 1·4.

Spectroheliograph. An instrument used for photographing the Sun in the light of one particular wavelength only. If adapted for visual use, it is known as a *spectrohelioscope*.

Spectroscope. An instrument used to split up the light of a star, or other luminous object (Fig. 79). Equipment based upon the principle of the spectroscope has given us practically all our knowledge about the nature and composition of the stars and galaxies.

The actual splitting-up of the light into a *spectrum* is carried out by means of a *prism or a *diffraction grating. Basically, the idea is straightforward enough, but an astronomical spectroscope is extremely complex, and has to be used in conjunction with a telescope. Moreover, the telescope must be large; otherwise, there will be too little light for useful analysis. (Matters are much easier with the *Sun, where there is plenty of light available, and the *solar tower arrangement can be used.) Nowadays, of course, all the research is carried out photographically.

If an incandescent solid, liquid, or high-density gas is examined with a spectroscope, the result is a continuous rainbow, from red at the long-wave end to violet at the short-wave end; this is a *continuous spectrum*. A gas at lower density, however, produces no rainbow; it shows an *emission spectrum*, made up of isolated bright lines. Each line is due to one particular *element or group of elements, and no element can duplicate the lines of another. This makes it possible to tell which substances are giving out the light. In practice, identifications are not easy, since any one element may give out a vast number of lines.

The Sun's bright surface or *photosphere gives a continuous spectrum. Overlying the photosphere is the *chromosphere, or solar atmosphere, which is made up of gas at a much lower density, and which should therefore yield a bright line or emission spectrum. Actually, the brilliant rainbow in the background causes the lines to appear dark, and the final result is an *absorption spectrum*, first studied in detail by J. von Fraunhofer

157

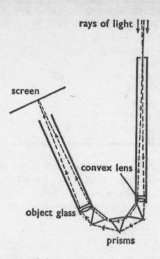

Fig. 79. *Spectroscope principle*

a century and a half ago; the dark lines are still known as
*Fraunhofer Lines. The positions of the lines are not affected,
so that identifications can be made. One example will show what
is meant. By itself, sodium vapour will produce two bright yellow
lines. In the spectrum of the Sun, there are two dark lines in the
yellow part of the rainbow, agreeing in position with the sodium
lines produced by laboratory experiments – and thus we can be
quite sure that there is sodium in the Sun.

Ordinary *stars show spectra of the same general type,
though the details differ; hot white stars have spectra which
are easily distinguished from those of cooler red stars. Gaseous
*nebulæ yield emission spectra, and so do *quasars. The spectra
of external *galaxies are somewhat confused, because they are
made up of the combined spectra of many millions of stars and
other objects, but the main absorption lines can still be identified.

Comparing the spectra of different stars makes it possible to
work out the comparative luminosities of the stars concerned,
and this in turn gives a clue as to their distances from us. Measures
of this kind are known as *spectroscopic parallaxes*, but the term
is not really a good one, because there is no to-and-fro shifting
in position as with ordinary trigonometrical *parallax.

Another vitally important branch of spectroscopic research

is concerned with the towards-or-away *radial velocities of the stars and galaxies. As has been described under the heading *Doppler Effect, an approaching body will have its spectral lines moved over to the short-wave or violet end of the spectrum, while a receding body will show a *red shift. All our ideas about the expansion of the universe depend upon this principle. In the unlikely event of our finding that the red shifts in the spectra of galaxies are not due to Doppler Effects, we shall have to do some very drastic rethinking!

So far as the *Moon and *planets are concerned, the spectroscope is rather less effective, because these bodies shine by reflected sunlight and have no inherent luminosity. However, a good deal of information about planetary atmospheres was obtained well before the age of space-probes. Thus Jupiter and the other giants were known to be rich in hydrogen, and the outer part of the atmosphere of Venus was found to be mainly carbon dioxide. Nowadays, of course, equipment carried on space vehicles has improved our knowledge beyond all recognition.

Spectroscopic Binary. A *binary star whose components are too close together to be seen separately. If they are in fairly quick motion round their common centre of gravity, one component will be approaching us while the other is receding (Fig. 80); the first component will show a violet shift in its spectrum, due to the *Doppler Effect, while the receding star will show a red shift. Consequently, the dark lines seen in the spectroscope will appear double. If only one star has a spectrum bright enough to be seen, then the lines will oscillate to and fro around a mean position. Many spectroscopic binaries are now known.

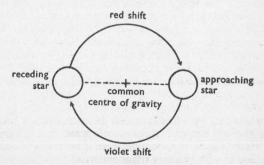

Fig. 80. *Spectroscopic binary*

Spectroscopic Parallax. This is described under the heading *Spectroscope.

Speculum. The main mirror of a *reflector. Older mirrors were made of a substance known as speculum metal; modern ones are of glass. For instance, the mirror of the new 98-inch reflector at the Royal Greenwich Observatory is made of Pyrex.

Spherical Aberration. The blurred appearance of an image as seen in a telescope, due to the fact that the lens or mirror does not bring the rays falling on its edge and its centre to exactly the same focus. If the spherical aberration is noticeable, then the lens or mirror is of poor quality, and should be corrected.

Spicules. Jets, up to 10,000 miles in diameter, in the Sun's *chromosphere. They last for only four to five minutes each, and are probably associated with the solar *granules. Spicules may be observed by means of a *Lyot filter, but are not visible in ordinary telescopes.

Spörer's Law. This is described under the heading *Sun.

Sputnik. The original Russian term for an artificial satellite. The first of all the artificial satellites, launched by Soviet scientists on 1957 October 4, was known as Sputnik I.

Star. A self-luminous gaseous body. It may also be defined as 'a sun', since the Sun is a normal star.

The stars are grouped into *constellations, and are lettered and numbered accordingly. For the most conspicuous stars, Greek letters are used, the brightest star in any constellation being the Alpha of that constellation, the second brightest star being Beta, and so on (though the strict sequence is often not properly followed – as in the constellation Sagittarius, where the two brightest stars are not Alpha and Beta, but Epsilon and Sigma!) For fainter stars, catalogue numbers are used.

The Greek alphabet is as follows:

α Alpha	η Eta	ν Nu	τ Tau
β Beta	θ Theta	ξ Xi	υ Upsilon
γ Gamma	ι Iota	ο Omicron	φ Phi
δ Delta	κ Kappa	π Pi	χ Chi
ε Epsilon	λ Lambda	ρ Rho	ψ Psi
ζ Zeta	μ Mu	σ Sigma	ω Omega

160

Many of the stars have individual or *proper names*, but in general these names are now used only with the stars of the first magnitude.

The *distances* of the stars are very great, and even Proxima Centauri, our nearest stellar neighbour (not counting the Sun, of course) is more than 4 *light-years away. The distances of the closer stars are measured by means of *parallax; with more remote stars, less direct methods have to be used. The six closest stars are as follows:

Star	Distance (light-yrs.)	Annual proper motion (sec. of arc)	Magnitude Apparent	Absolute	Luminosity (Sun = 1)
Proxima Centauri	4·2	3·75	10·5	15·1	0·0001
α Centauri	4·3	3·61	−0·3	4·4+5·8	1·1 +0·2
Barnard's Star	6·0	10·27	9·5	13·2	0·0005
Wolf 359	8·1	3·84	13·5	16·5	0·00002
Lalande 21185	8·2	4·75	7·5	10·5	0·005
Sirius	8·7	1·21	−1·4	1·3	26

α Centauri is a fine binary. Sirius has a dim White Dwarf companion. All the stars in the list, except for Sirius and α Centauri, are Red Dwarfs of spectral type M.

Most of the naked-eye stars are much further away than this; for instance, the distance of the Pole Star is approximately 700 light-years. The ten apparently brightest stars are as follows:

Star	Proper name	Apparent magnitude	Spectrum	Distance (light-years)	Luminosity (Sun = 1)
α Canis Majoris	Sirius	−1·4	A	8·7	26
α Carinæ	Canopus	−0·7	F	Over 600	80,000
α Centauri	—	−0·3	G+K	4·3	1·1 +0·2
α Boötis	Arcturus	−0·1	K	41	100
α Lyræ	Vega	0·0	A	26	50
α Aurigæ	Capella	0·0	G	47	150
β Orionis	Rigel	0·1	B	About 900	50,000
α Canis Minoris	Procyon	0·4	F	11	5
α Eridani	Achernar	0·5	B	66	200
α Orionis	Betelgeux	Variable	M	190	1200

α Centauri is a binary; Sirius and Procyon have White Dwarf companions, and Capella is an extremely close binary. α Centauri, Canopus and Achernar are too far south to be seen in Europe. Both Canopus and Rigel are so remote that the values given for their distances and luminosities are uncertain.

No telescope will show a star as anything but a point of light, and so our knowledge of them has to be obtained mainly by instruments based upon the principle of the *spectroscope. Hot stars show spectra which are decidedly different from cool stars

– and the range is considerable; some white stars have surface temperatures of over 100,000 degrees Centigrade, while at the other end of the scale a few curious red stars have been found whose surface temperatures seem to be only about 500 to 600 degrees Centigrade. The Sun is a very normal star, with a surface temperature of 6000 degrees Centigrade. One of the most powerful stars known is S Doradûs, in the large cloud of Magellan, which is a million times as luminous as the Sun. At the other end of the scale comes the dim red Ross 614B with only 1/2300 of the Sun's luminosity.

The stars are divided into 10 spectral types, each of which is again subdivided. The main types are as follows:

Type	Surface temperature (degrees C.)	Colour	Typical star	Remarks
W	Over 36,000	Greenish white	γ Velorum	Wolf-Rayet stars, with emission lines in spectra
O			ζ Puppis	Helium lines prominent
B	28,600	Bluish	Spica	Helium lines prominent
A	10,700	White	Sirius	Hydrogen lines prominent
F	7500	Yellowish	β Cassiopeiæ	Calcium lines prominent
G	giant 5200	Yellow	ε Leonis	Metallic lines numerous
	dwarf 6000		The Sun	
K	giant 4230	Orange	Arcturus	Hydrocarbon bands appear
	dwarf 4910		ε Eridani	
M	giant 3400	Orange-red	Betelgeux	Complex spectra with many bands. Many M-type giants are variable
	dwarf 3400		Wolf 359	
R	2300	Orange-red	U Cygni	Carbon bands
N	2600	Very red	S Cephei	Carbon bands
S	2600	Red	R Andromedæ	Zirconium oxide bands. Mostly long-period variables

Near a star's centre, the temperature rises to a fantastic value – around 14,000,000 degrees Centigrade in the case of the Sun, which is rather mild by stellar standards. In these regions, where the star is producing its energy, the atoms are of course completely broken up.

The *source of stellar energy* has been carefully studied, and it is now certain that the cause lies in nuclear reactions. Broadly speaking, a star such as the Sun is changing its hydrogen into helium, releasing energy and losing mass as it does so. The Sun is losing mass at the rate of 4,000,000 tons per second, and hotter, more massive stars such as Rigel are suffering even greater mass-losses. Though the Sun will not change much for at least 5000 million years in the future, Rigel and its kind can hardly remain brilliant for more than a few millions of years, so that on the cosmical time-scale they are short-lived.

We have also learned a good deal about *stellar evolution*, and though we do not yet know the full life-story of a star we are modestly confident of being on the right track. It seems that a star begins its career inside a *nebula. At first it is nothing more than a local condensation in the excessively rarefied material, but as the embryo star shrinks, under the influence of gravitation, its interior becomes hotter and hotter. Subsequent events depend almost entirely upon the initial mass of the star. If the mass is less than about one-tenth that of the Sun, nuclear reactions will never begin, and the star will remain of low luminosity; eventually it will turn into a dead globe – a black dwarf, in fact. If the mass is greater, then nuclear reactions will start as soon as the core temperature has risen sufficiently, and the star will join the *Main Sequence of the *Hertzsprung-Russell Diagram, where it will remain for a long period. The Sun is now in this stage of its evolution, and, as we have seen, is producing its energy by the conversion of hydrogen into helium.

When the supply of available hydrogen is exhausted, the star has to change its structure drastically. The outer layers expand and cool, while the helium-rich core shrinks; the star becomes a Red Giant, to the upper right of the H-R Diagram. When the Sun reaches this stage, it will have a diameter of perhaps 25,000,000 miles and a luminosity 100 times its present value – with disastrous results for our Earth!

The further contraction of the core means a new rise in temperature, until the helium starts to react in its turn; this is the so-called 'helium flash'. Heavier and heavier elements are produced, following a series of highly complicated reactions, until all the available nuclear energy has been spent. The star then collapses, moving to the lower left of the H-R Diagram and becoming a *White Dwarf; the density may attain over 200,000 times that of water. The atoms making up the star are crushed together, with little waste of space, which accounts for the high density. After another immensely long period, the star loses the last of its light and heat.

If the original star is more than about 3½ times as massive as the Sun, it suffers a different fate. The nuclear reactions finally become 'out of control', so to speak, and the star becomes a

163

*supernova; it explodes, hurling most of its material away into space, and leaving the remnant of the original star as a small, incredibly dense body made up of neutrons. The density is even greater than with a White Dwarf, because even the protons and electrons of the original material are forced together. A matchboxful of *neutron star material could well weigh at least a million tons. Supernova remnants are seen as clouds of expanding gas, and the remnants of the original stars send out quickly-varying radio emissions, so that they are known as *pulsars.

If the star is more than about ten times as massive as the Sun, it may continue to collapse indefinitely when the main nuclear reactions cease – in which case it will become a *black hole, though it must be stressed that our knowledge of this part of the career of a very massive star is still very meagre, and the existence of black holes has yet to be definitely proved.

This account of stellar evolution is very incomplete and oversimplified, but may serve to give a general picture. We know, at any rate, that no star can last for ever. Dead stars are presumably common in the Galaxy, though since they emit no light or heat we cannot detect them.

The stars are of many kinds, and even a small telescope will show many interesting *binaries and *variables. Altogether there are about 100,000 million stars in our *Galaxy – and we must always remember that even the Galaxy is only a very small part of the universe.

Star of Bethlehem. The star mentioned in the Gospel according to St Matthew (Chapter 2) as having led the Wise Men from the East to the place of the Nativity. Many theories to explain it have been proposed – the favourite being a conjunction of two or more planets, appearing so close together that they merged into one exceptionally brilliant object. Unfortunately, it has been shown that no such conjunctions were visible from the Holy Land at this time, and the idea is definitely untenable. It is not likely that the Star can have been a nova or supernova, as there is no mention of anything of the kind by contemporary astronomers, and in any case an outburst of this kind would remain visible for many nights. *Halley's Comet returned in BC 11, but this too would have been on view for some time, and does not seem to be a likely candidate. Neither can the Star have been a familiar body (such as Venus), as it would have caused no excitement whatsoever! As we have nothing to guide us apart from the brief mention by St Matthew, we must admit that there is no scientific explanation for the Star of Bethlehem.

Steady-State Theory. This is described briefly under the heading *Universe.

Stratosphere. The layer in the Earth's *atmosphere lying above the *troposphere. It extends from about 7 to about 40 miles above sea-level.

Struve, F. G. W. (1793–1864). German astronomer who spent much of his career in Russia as Director of the *Pulkova Observatory. He was the real founder of the astronomy of double stars.

Struve, Otto (1819–1905). Son of *F. G. W. Struve, who succeeded him as Director at *Pulkova and was also a great double-star observer.

Struve, Otto (1897–1963). Great-grandson of *F. G. W. Struve. He was essentially a stellar spectroscopist; one-time Director of the *Yerkes Observatory and subsequently of the radio astronomy observatory at Green Bank, West Virginia.

Sun. The star which is the central body of the *Solar System, and around which the planets revolve. Its mean distance from us is 92,957,209 miles, which is not very far by cosmical standards. The diameter is 865,000 miles; the volume is 1,300,000 and the mass 333,000 times that of the Earth (Fig. 81), and the *specific gravity is 1·4.

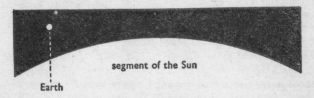

segment of the Sun

Earth

Fig. 81. *Size of the Sun compared with the Earth*

The Sun is a *Main Sequence star of spectral type G, and is producing its energy by the conversion of hydrogen into helium, mainly by the so-called *proton-proton reaction. It is losing mass at the rate of 4,000,000 tons per second, but it is in no

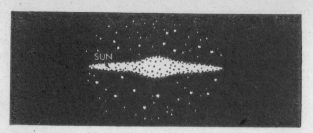

Fig. 82. *Shape of the Galaxy, showing the position of the Sun*

danger of imminent extinction; it is at least 5000 million years old, and is not likely to alter much for 5000 million years or so in the future.

It lies well away from the centre of the *Galaxy, on the inner edge of a spira' arm (Fig. 82). The distance between the Sun and the galactic centre is thought to be about 32,000 light-years. Of course, the Sun shares in the general rotation of the Galaxy; its velocity is about 135 miles per second, so that it takes 225,000,000 years to complete one circuit – a period which has been unofficially termed the 'cosmic year'.

Telescopically, the Sun often shows dark patches known as *sunspots*. The spots look blackish, but are not really so; their temperature is around 4000 degrees Centigrade, as against 6000 degrees for the main bright surface or *photosphere. A large spot consists of a dark central part or *umbra, surrounded by a lighter *penumbra (Fig. 83), though small spots are often lacking in penumbra. It is also clear that spots tend to appear in groups, some of which are highly complex – though most major groups have two principal spots, a 'leader' and a 'follower'.

umbra   penumbra

Fig. 83. *Structure of sunspots*

166

Spot-groups may be of immense size, utterly dwarfing the Earth, and they are sometimes visible with the naked eye. They have many interesting features; for instance, they are associated with strong magnetic fields. Small spots may last for only a few hours, though larger groups have been known to persist for several months.

The Sun shows a semi-regular cycle of activity, with a period of about 11 years. At maximum activity, as in 1947, 1958 and 1969, the disk may contain many spot-groups at any one moment; at minimum, as in 1953 and 1964, the disk may be free of spots for weeks at a time (Fig. 84). The first spots of a new cycle appear in moderately high latitudes on the Sun; as the cycle progresses, the spots approach nearer and nearer the equator,

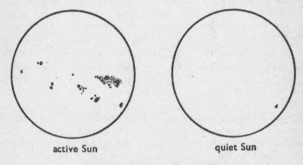

active Sun   quiet Sun

Fig. 84. *Active and quiet Sun*

while the first spots of a new cycle may appear before the last spots of the old cycle have died away. This behaviour is known as *Spörer's Law, in honour of its discoverer. We must admit that so far the reasons for the solar cycle and Spörer's Law are not well understood.

Spots are associated with the bright irregular *faculæ, which may be regarded as luminous 'clouds'. Faculæ are visible in ordinary telescopes, but the Sun's atmosphere, made up of the *chromosphere and the *corona, is not; it can be seen with the naked eye only during a total solar *eclipse, though with instruments based on the principle of the *spectroscope the *chromosphere, together with the spectacular red *prominences, may be studied at any time (the corona is much more elusive). Mention should also be made of solar *flares, which are short-lived, violent outbursts, usually associated with active spot-groups.

As well as sending us visible light, the Sun is a source of

radio waves; these long-wavelength emissions may be picked up with relatively simple equipment. Radio studies of the Sun have provided invaluable information during the past twenty years. The Sun is also an X-ray source.

It is highly dangerous to look at the Sun through a telescope, even when a dark glass is placed over the eyepiece. The only sensible way to observe sunspots is to use the telescope to project the Sun's image on to a white screen. In this way, the spots may be followed with no risk to the observer's eyesight, and their apparent drifts may be noted. The Sun takes several weeks to rotate upon its axis (25·4 days at the solar equator, 29 to 30 days near the poles) and so the spots are carried across the disk from one side to the other. When a spot vanishes over the limb, it may be expected to reappear at the opposite limb just over a fortnight later, provided that it still exists. Spots seen close to the limb are naturally foreshortened.

Solar physics is a most important branch of modern astronomy, and many observatories concentrate entirely upon it. The Sun is a close neighbour of ours, and by studying it we are also learning about the other stars, which we can never see as anything but points of light.

Sundial. An instrument used to show the time, by using an inclined rod or plate (the *gnomon or *style*) to cast a shadow on to a graduated dial. The gnomon points to the celestial pole. The sundial shows apparent time; to obtain mean time, the value shown on the dial must be corrected by allowing for the *equation of time.

Supergiant Stars. Stars of exceptionally great luminosity and low density.

Superior Planets. The planets beyond the orbit of the Earth in the Solar System – that is to say, all the principal planets apart from Mercury and Venus.

Supernova. A star which 'explodes', sending much of its material away into space and never returning to its old form. For a brief period supernovæ send out tremendous amounts of energy, and so they can be seen over vast distances.

With a star more than about $3\frac{1}{2}$ times as massive as the Sun, all kinds of complicated processes come into operation when the star has left the *Main Sequence. If the helium core becomes too massive, the whole delicate balance of the star is upset, and the core temperature rises to fantastic values; we must also consider *neutrinos, which can carry off a great deal of energy. When the internal temperature has soared to some 5000 million degrees, a sudden change occurs. The heaviest nuclei produced up to this time are those of iron, but then the iron nuclei split up and are converted back to helium; gravitational forces result in a sudden collapse of the core, and the outer layers of the star are abruptly heated to about 300 million degrees. The result is a supernova outburst. When the cataclysm is over, we are left with an expanding gas-cloud; the remnant of the original star is made up of *neutrons, caused by the running-together of protons and electrons. Massive though it may be, a neutron star can be no more than about 100 miles in diameter.

Supernovæ are of two main types. The more luminous (Type I) have very little hydrogen in their outer layers, while Type II supernovæ have a good deal. Unfortunately, no supernova has been seen in our own Galaxy since 1604, which was before the invention of the telescope. Previous supernovæ had been recorded in 1006 (in Lupus), 1054 (in Taurus), and in 1572 (Tycho's Star, in Cassiopeia). All these have become brilliant enough to be seen with the naked eye in broad daylight. All can be identified today as radio sources; the 1054 star has produced the *Crab Nebula. It is thought that another supernova occurred in about 1702, in Cassiopeia, at a distance of around 11,000 light-years, but was not observed because it was hidden by interstellar material. Many galactic radio sources are also thought to be due to supernovæ which blazed forth in prehistoric times.

On average, each galaxy of our kind seems to produce one supernova in 600 years or so, so that on this reckoning we are not yet due for another. Of course, there is no means of telling, and a supernova might be seen at any time. Astronomers hope so, since they would welcome the chance to study one with modern equipment. Meanwhile, many supernovæ have been observed in external galaxies, and one of them – the 1885 supernova in the Andromeda Galaxy, M.31 – was just visible to the naked eye.

Sutherland. The new observing centre for astronomy in South Africa; the largest telescope is the 74-inch reflector formerly at the Radcliffe Observatory in Pretoria. The present Director of the South African Astronomical Observatories is Sir Richard Woolley, former Astronomer Royal.

Synodic Period. The interval between successive *oppositions of a *superior planet. The periods for each planet are given on page 125. Mars has much the longest synodic period, because on the astronomical scale it is not far beyond the Earth's orbit, and moves at a comparable speed. With regard to an *inferior planet, the term is applied to the interval between successive conjunctions with the Sun.

Syrtis Major. The principal dark marking on *Mars; formerly known as the Kaiser Sea or Hourglass Sea. It has a V-shape. It is now known to be a plateau rather than a depression. Under suitable conditions it is visible with a small telescope.

Syzygy. The position of the Moon in its orbit when at new or full.

T

Tektites. Small, glassy objects found in a few restricted areas on the Earth, notably in Australia. They are decidedly mysterious, since nobody knows where they came from; they may or may not be of extra-terrestrial origin. The main European tektite field is in Czechoslovakia. No tektite has ever been found in the British Isles.

Telescope. The main instrument used to collect the light from celestial bodies. There are two main types, the *reflector and the *refractor. The first refractor was made in Holland in or about 1608, while the first reflector was constructed by Sir Isaac Newton and submitted to the Royal Society in 1671.

Many large refractors were constructed during the latter part of the 19th century; pride of place goes to the Yerkes telescope, which has an object-glass 40 inches in diameter. Today, however, the main emphasis is on reflectors, since a large mirror is easier to make and use than a large lens. The Palomar 200-inch reflector has been in operation for over a quarter of a century; the 236-inch reflector made in the USSR and set up in Siberia is still being tested. There are now several reflectors with apertures of between 98 and 160 inches, and more are being planned.

Telescopes have also been sent into space, and within the next few decades it is likely that there will be a full-scale observatory on the Moon – which will be an ideal site, because of the lack of atmosphere.

Temporary Stars. An old term for *novæ.

Terminator. The boundary between the day and night hemispheres of the Moon or a planet (Fig. 85). Since the lunar surface is mountainous, the terminator is rough and jagged, and isolated peaks may even appear to be detached from the main body of the Moon. Mercury and Venus, which also show lunar-type phases, seem to have smooth terminators, though Mercury is now known to have a rough surface, just as rough as that of the Moon. Mars, which may be appreciably *gibbous when well away from *opposition, also shows a smooth terminator, no doubt for the same reason.

Tethys. The third *satellite of Saturn.

Themis. A reported *satellite of Saturn, described by Pickering in 1904, and said to move between the orbits of *Titan and *Hyperion. It has never been confirmed, and apparently does not exist.

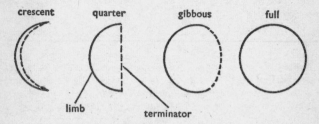

Fig. 85. *Limb and terminator*

Thermocouple. An instrument used for measuring very small quantities of heat. Basically, it consists of a circuit made up of wires of two different metals, joined together. If one of the joins is warmed, and the other kept at a constant temperature, an electric current is set up in the wire; the amount of the current gives a key to the amount of heat involved. Using thermocouples, together with large telescopes, remarkably feeble heat-sources may be detected.

Tides. The daily rise and fall of the ocean waters, due to the gravitational pulls of the Moon and (to a lesser extent) the Sun.

If the Earth were surrounded by a uniform shell of water the Moon would tend to heap up the water beneath it; as the Earth span on its axis, this 'heap' would stay under the Moon, and so would sweep right round the Earth. Since there would also be a 'heap' on the far side of the Earth, each place would have two high tides per day. Actually, matters are not so straightforward as this, because the oceans are irregular both in outline and in depth, but the general principles are easy to understand.

When the Sun and Moon are pulling in the same direction, with the Moon at *syzygy, the tides are relatively strong (spring tides, as in top diagram, Fig. 86). When the Moon is at *quadrature (bottom diagram, Fig. 86), it is pulling against the Sun, and the tides are weaker (*neap tides*).

172

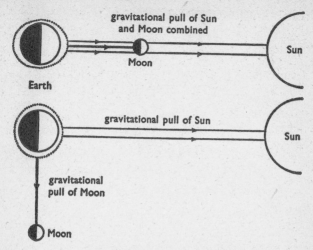

Fig. 86. *Spring and Neap tides*

Time. See under the headings *Sidereal time, *Solar time, *Greenwich Mean Time, *Universal time, and *Equation of time.

Titan. The largest satellite of Saturn. Details of its orbit and dimensions are given in the table on page 146. Estimates of its diameter range between 3500 and 2600 miles; at any rate, Titan is considerably larger than the Moon – and it is the only satellite in the Solar System known to have an atmosphere of appreciable density. The ground pressure may be as high as 100 millibars, ten times as great as the pressure at the surface of Mars, and it is quite likely that there are clouds, though from Earth it is naturally very difficult to see any surface details.

Titan has a low escape velocity, and must regularly lose atoms and molecules from its atmosphere; but these particles cannot escape from the pull of Saturn itself, and so Titan 're-collects' them as it moves round the planet. In fact, the atmosphere of Titan is constantly recycled. Methane is an important constituent, but for full information we must await the results of the Saturn probes which should by-pass the planet in 1979. Titan is of such interest to astronomers that there is serious talk of dispatching a probe primarily to study it.

Titan was discovered by the Dutch astronomer Huygens in 1655, and is bright enough to be seen with a small telescope.

Titania. The third *satellite of Uranus.

T.L.P. Transient Lunar Phenomenon. Over many years, serious lunar observers have reported elusive local obscurations and reddish glows in certain areas of the Moon, notably near the brilliant crater Aristarchus. These are now known as T.L.P.s (a term for which I believe I was responsible!). Until 1958 the reports were treated with considerable scepticism, but then a T.L.P. in the crater Alphonsus was observed by the Russian professional astronomer N. A. Kozyrev, at the Crimean Astrophysical Observatory, and confirmatory spectrograms were obtained. Further reliable observations have been made since then, and the existence of T.L.P.s is no longer seriously questioned, though interpretations differ; it may be that the phenomena are due to gaseous emissions from below the Moon's crust.

There is a definite link between T.L.P.s and the mild ground tremors or 'moonquakes' recorded by the instruments set up on the Moon by the Apollo astronauts, so that the phenomena are definitely of internal origin. They are, of course, very weak by terrestrial standards, and certainly cannot be classed as truly volcanic outbreaks.

Tower Telescope. Equipment used for studying the Sun. Its chief advantage is that the heavy spectrographic equipment need not be moved at all, since the sunlight is reflected to a fixed point by means of a *cœlostat fitted at the top of a tower.

Transfer Orbit. The most economical path or orbit for a spacecraft launched toward another planet. To move in a 'straight line' would need far too much fuel. Instead, the probe is put into an orbit which will take it out in an ellipse to the orbit of the target planet, so that the planet and the probe will meet. Almost all the journey is therefore carried out in free fall, without the expenditure of fuel.

Transit. The passage of a celestial body across the observer's *meridian. Thus the *First Point of Aries must transit, or cross the meridian, at 0 hours sidereal time.

There are several other astronomical meanings of the word. Mercury and Venus are said to transit when they pass between the Earth and the Sun, so that the planet shows up as a black spot against the solar surface; transits of Mercury (Fig. 87) are not particularly uncommon (the next will take place on 1986 November 12), but the next transit of Venus is not due until the

Fig. 87. *Transits of Mercury*

year 2004. Similarly, a satellite of a planet is in transit when seen against the planet's disk. Transits of the four large satellites of Jupiter are easy to see even with a small telescope.

As Jupiter spins on its axis, various markings, such as spots, are brought to the planet's central meridian. When a marking reaches the central meridian, as shown in the diagram (Fig. 88), it is in transit.

Fig. 88. *Central meridian of Jupiter*

Transit Instrument. A telescope specially mounted, so that it can be used for timing the *transits of stars across the *meridian. It moves only in *declination, and always points to the meridian. When a star approaches the meridian, it may be seen in the telescope field, and it crosses a series of wires set up at the focus of the object-glass (Fig. 89), so that the moment of its transit may be timed very accurately. Transit instruments were once the basis of all precise time-keeping, though by now they have been largely superseded.

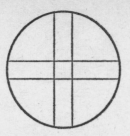

Fig. 89. *Cross-wires*

The transit instrument set up at Greenwich Observatory by Sir George Airy, Astronomer Royal for many years during the 19th century, is taken to mark the Earth's prime meridian (longitude 0 degrees).

Triton. The larger *satellite of Neptune. It is comparable in size with Mercury, and is not a difficult object; it was discovered by Lassell not long after the discovery of Neptune itself. It is unique among large satellites inasmuch as it has retrograde motion.

Trojans. The Trojans are *minor planets which move in almost the same orbit as Jupiter (Fig. 90). There is no danger of their colliding with Jupiter, since one group of Trojans keeps well ahead (by about 60 degrees), while the other group is equally far behind. More than a dozen Trojans are known, but all are faint,

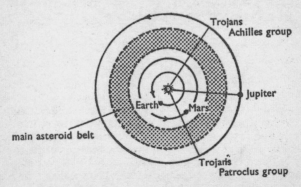

Fig. 90. *The two groups of Trojan asteroids*

because they are so far from us. Each is named after a hero of the legendary Trojan War; the two brightest members of the group, Achilles and Patroclus, are over 100 miles in diameter. The other Trojans of magnitude 16 or brighter are Hector, Nestor, Priam, Agamemnon, Odysseus, Æneas, Anchises, Troilus, Ajax, Diomedes and Antilochus.

Tropical Year. This is described under the heading *Year.

Troposphere. The lowest part of the Earth's atmosphere, reaching to a height of about 7 miles on an average. It includes most of the mass of the atmosphere, and all normal clouds lie within it. Above it, separating it from the *stratosphere, is a shallow layer known as the *tropopause*.

T Tauri Stars. Very young stars, of low luminosity; they vary irregularly in light, and are often associated with nebular material. T Tauri itself is the prototype star.

Twilight, Astronomical. The state of illumination when the Sun is below the horizon, but by less than 18 degrees. In Britain, near midsummer, twilight lasts all night.

Twinkling. The popular name for *scintillation.

U

UFOs. See *Flying Saucers.

U Geminorum Stars. Variable stars, which show periodical outbursts. The prototype star is U Geminorum; the brightest member of the class is SS Cygni, which can rise to about the 8th magnitude.

Ultra-Violet Radiation. Electromagnetic radiation which has a wavelength shorter than that of violet light, and so cannot be seen with the human eye. Still shorter wavelengths are termed *X-rays*.

Umbra. The main cone of shadow cast by the Earth. Further details are given under the heading *Eclipses, Lunar.

Umbriel. The second *satellite of Uranus.

Universal Time. The same as *Greenwich Mean Time.

Universe. Space, together with all the matter and energy contained in it – in fact, everything that exists! Whether it is finite or infinite, we do not know. Problems of this sort seem indeed to be too difficult for our limited intellects to solve. If space is limited, we cannot imagine what could lie outside it; if unlimited, we are faced with the task of picturing something which goes on literally for ever. To say that space is 'finite but unbounded' does not make things much easier.

If the *red shifts in the spectra of the *galaxies are genuine *Doppler Effects, then we must assume that the universe is in a state of expansion, but we know nothing definite about its origin, and all we can do is to work from some arbitrarily-chosen starting-point. On one theory – the *steady-state theory*, put forward in 1948 by H. Bondi and T. Gold and later developed by F. Hoyle and others – the universe has always existed, and will exist for ever; so that as old galaxies die, new material is continuously created out of nothingness, in the form of hydrogen atoms. The idea is ingenious, but it is in no way a solution of the problem of the actual creation of the universe, and in any case it has so many weak links that most astronomers have now rejected it.

The older *evolutionary theory* supposed that all the matter in the universe was created at one moment, more than 10,000 million years ago. This 'primæval atom', as it has been termed, exploded and sent material outward in all directions; there may have been a temporary state of balance, but then the expansion began again, and is still going on now that the material has been formed into galaxies. Here too there are serious difficulties, and nobody has yet been able to explain just how the primaeval atom came into existence. Moreover, it is hopeless to try to form any idea of the actual 'start of time'.

It may well be that the universe is in an oscillating condition, with alternate expansion and contraction; if so, we are living in a phase of expansion. But when all is said and done, our ignorance of the earliest period in the history of the universe is virtually complete, and neither can we tell if or when the universe will come to an end.

Uraniborg. Tycho *Brahe's observatory on the Baltic island of Hven. Tycho built it, and used it from 1576 to 1596; it was then abandoned, and nothing now remains of it.

Uranus. The seventh planet in order of distance from the Sun. Details of its globe and orbit are given under the heading *Planets.

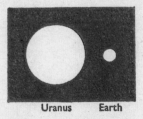

Uranus   Earth

Fig. 91. *Comparative sizes of Uranus and Earth*

Uranus was first identified in 1781 by William Herschel; it had been seen earlier by various observers, but had always been mistaken for a star. It is just visible to the naked eye under good conditions, and a telescope shows its pale, greenish disk, though few surface details are to be made out. It is a giant, over 29,000 miles in diameter, and therefore much larger than the Earth

179

(Fig. 91); in composition it is of the same basic type as the other giant, Neptune.

Its main peculiarity is in the tilt of its axis, which amounts to 98 degrees (Fig. 92). Since this is more than a right angle, the rotation of Uranus is technically retrograde, though it is not generally regarded as such. The 'seasons' there would be most remarkable, but it is obvious that Uranus is not a world capable of supporting any form of life as we know it. None of the five satellites can be seen except with a moderately large telescope.

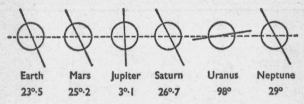

Earth	Mars	Jupiter	Saturn	Uranus	Neptune
23°·5	25°·2	3°·1	26°·7	98°	29°

Fig. 92. *Axial inclinations of the planets*

Ursa Major. The most famous northern constellation, *circumpolar in Britain. Its seven main stars make up the pattern known commonly as the Plough or (in America) the Big Dipper.

UV Ceti Stars. *Flare stars. UV Ceti is the most famous member of the class.

V

Van Allen Zones (or Van Allen Belts). Zones around the Earth in which electrically-charged particles are trapped and accelerated by the Earth's magnetic field (Fig. 93). They were discovered in 1958 by J. van Allen and his colleagues in the United States, from results obtained with the first successful American artificial satellite, Explorer I. Their existence had been previously unsuspected, and their detection came as a considerable surprise.

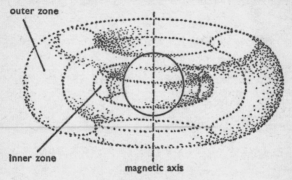

Fig. 93. *The Van Allen zones*

There seem to be two main belts. The outer one, made up chiefly of *electrons, is very variable, since it is affected by events taking place in the Sun; the inner zone, composed mainly of *protons, is more stable.

Variable Stars. Most stars are constant in brightness, remaining unchanged for year after year, century after century. Some, however, brighten and fade over much shorter periods, ranging from a few hours up to a few years. These are the *variable stars*.

Eclipsing variables, or eclipsing binaries, are not genuinely variable at all. In such a system, there are two stars moving around their common centre of gravity; if one star passes in front of the other as seen from Earth, it causes an eclipse (or, more accurately, an occultation), and the combined light of the pair fades until the eclipse is over. Of course, the components are much too close together to be seen separately. The most famous star of this kind is Algol in Perseus; one component is appreciably brighter than the other, so that every 2½ days, when the brighter component is partly eclipsed, Algol fades down from magnitude 2 to magnitude 3½, remaining at minimum for about twenty minutes before starting to recover its light. With Beta Lyræ, near Vega, the components are more equal; the star always seems to be in variation, and the system is a remarkable one, since spectroscopic investigations have shown that both components are drawn out into egg-like shapes.

Cepheids have shorter periods, and are perfectly regular. Like all true variables, their light-changes are due to real changes in luminosity. They have been described under a separate heading, as have the *RR Lyræ stars, which were formerly known as cluster-Cepheids.

Long-period variables are usually (but not always) Red Giants of late spectral type – that is to say, types M, R, N or S. They have large ranges of magnitude, and periods of from several weeks up to well over a year. Mira Ceti, the most celebrated member of the class, may reach magnitude 2, and has been known to outshine the Pole Star, but at other maxima it may not even reach the third magnitude, while at minimum it becomes so faint (magnitude 9) that even binoculars will not show it. Its mean period is 331 days, but, like all long-period variables, it is not perfectly regular in its behaviour.

Irregular variables have no periods at all, and are always likely to catch observers by surprise. With stars such as Betelgeux in Orion, the changes are relatively slow and slight; these are termed *semi-regular* variables. True irregular variables are much more erratic. There is probably a close link between some of the irregular variables and the explosive stars known as *novæ.

The study of variable stars makes up an important branch of modern stellar research. With the long-period and irregular stars, amateurs with moderate telescopes can do valuable work; the method of observation is to compare the variable with nearby stars of known and constant brilliancy, so that the magnitude of the variable may be worked out and a *light-curve drawn (Fig. 94).

Variation. An inequality in the motion of the Moon, due to the fact that the Sun's pull upon it throughout its orbit is not constant in strength.

The term is also used with respect to the direction shown by a magnetic compass on the Earth. A compass needle will point to

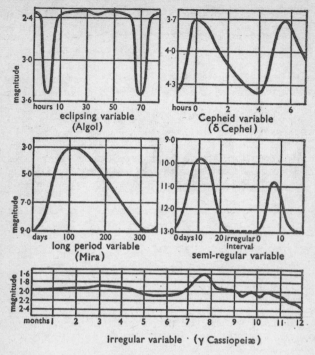

Fig. 94. *Light-curves of variable stars*

the magnetic pole, not to the geographical pole – which is some distance away. Magnetic variation is the difference, in degrees, between true north and magnetic north.

Venus. The second planet in order of distance from the Sun. Details of the orbit and globe are given under the heading *Planets.

When seen with the naked eye, Venus is remarkably beautiful, and is much brighter than any other planet or star; at its best it can even cast a shadow. It is popularly known as the Evening Star or Morning Star, because it appears in the western sky after sunset or in the east before dawn; it can never be seen throughout a night, but may at times rise 5 hours before or set 5 hours after the Sun. Keen-sighted persons can catch sight of it in broad daylight, provided that they know where to look.

Since Venus is almost as large as the Earth, and is closer to us than any other planet, it might be expected to show considerable surface detail when observed with a powerful telescope. In fact, it does not, because its true surface is permanently hidden by its dense, 'cloudy' atmosphere. Generally, nothing is visible except for the characteristic *phase (Fig. 95), and the vague markings glimpsed from time to time are purely atmospheric in nature.

Fig. 95. *Phases of Venus, showing the changing apparent diameter. All drawings are to the same scale*

Before 1962 Venus was regarded as a planet of mystery. Its rotation period was unknown, and nobody could be sure whether the surface was water-covered or completely dry. Spectroscopic analysis had shown that the upper atmosphere was very rich in carbon dioxide, but for obvious reasons nothing was known about the lower layers. Then, however, the American probe Mariner 2 by-passed Venus within 21,000 miles, and showed that the surface temperature was much too high for water to exist in the liquid form (Fig. 96). Subsequently the Russians were able to soft-land several probes, and further information was obtained from another American fly-by, Mariner 5. In February 1974 Mariner 10 sent back the first close-range pictures of the upper cloud patterns. By then, radar results had indicated that Venus is covered, at least in part, with large and rather shallow craters.

It now seems that Venus is a decidedly unprepossessing world. The surface temperature is of the order of 900 degrees Fahrenheit, and the atmospheric pressure is about 100 times as great as that of the Earth's air at sea-level; carbon dioxide is the main constituent, but the upper clouds also contain materials such as sulphuric acid. The solid body of the planet rotates in a period of 243 days – longer than Venus's 'year' of 224·7 Earth-days – but the upper clouds spin round in only 4 days, so that clearly the

184

atmospheric structure is most peculiar. Venus rotates in a retrograde direction; the length of the 'solar day' on Venus is 117 Earth-days. Pictures obtained from the Russian Venereas 9 and 10 in 1975 show a rock-strewn surface.

Venus has a negligible magnetic field, and, like Mercury, is unattended by any satellite.

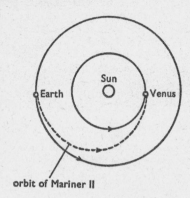

orbit of Mariner II

Fig. 96. *Orbit of Mariner II*

Vernal Equinox. The *First Point of Aries.

Vesta. The brightest, though not the largest, of the *Minor Planets. It was discovered by the German amateur H. Olbers in 1807, and is just visible to the naked eye when well placed.

Vogel, H. C. (1842–1907). Pioneer German stellar spectroscopist. It was he who established the existence of *spectroscopic binaries, in 1890.

Vulcan. During the 19th century it was believed that there must be a planet moving round the Sun at a mean distance much less than that of Mercury. The French astronomer, Le Verrier, whose mathematical calculations had led to the discovery of Neptune in 1846, was convinced of its reality, and it was even given a name – Vulcan. However, its existence was never confirmed, and it is now safe to assume that there is no planet within the orbit of Mercury.

185

W

White Dwarf. A very small, amazingly dense star. The atoms in it have been broken up, and the various parts packed tightly together with almost no waste space, so that the density rises to millions of times that of water; a spoonful of White Dwarf material would weigh many tons! Evidently a White Dwarf has used up all its nuclear 'fuel', and is shining feebly because it is still slowly shrinking. It is near the end of its active career, and has been aptly described as a 'bankrupt star'.

White Dwarfs are common in space, but they are so dim that they are hard to find unless they are relatively close to us. The best-known example is the companion of Sirius, which is smaller than the planet Uranus, but is almost as massive as the Sun. A more extreme case is Kuiper's Star, where the diameter is only about 4000 miles. If it were possible to take a cube of its material, with each side of the cube measuring one-tenth of an inch, the weight would be half a ton if measured under Earth conditions.

Widmanstätten Patterns. If an iron *meteorite is cut, polished and then etched with acid, characteristic figures of the iron crystals appear. These are known as Widmanstätten patterns, and are never found except in meteorites.

Wilson Effect. When a sunspot is close to the Sun's limb, it will be foreshortened, so that a spot which is actually circular will appear elliptical (Fig. 97). In 1769 the Scottish astronomer Wilson pointed out that in general, the *penumbra of the spot will appear narrower in the direction toward the Sun's centre than toward the direction of the limb, and he concluded that spots must be saucer-shaped depressions. If the spots were elevations, then the penumbra would appear narrower toward the limb.

This *Wilson Effect* seems to vary from spot to spot, and investigations are still going on, but it is generally thought that most spots are real depressions.

Wolf, Max (1863–1932). German pioneer of astronomical photography. He discovered almost 600 *minor planets.

Wolf-Rayet Stars. Exceptionally hot, greenish-white stars whose

Fig. 97. *Sunspots: the Wilson Effect*

spectra contain bright or emission lines as well as the usual dark absorption lines. Their surface temperatures may approach 100,000 degrees Centigrade, as against only 6000 degrees for the Sun, and they seem to be surrounded by rapidly-expanding envelopes of gas. Attention to them was first drawn by the two astronomers Wolf and Rayet in 1867. They are comparatively rare, and are included in the spectral types W and O.

W Virginis Variables. Short-period variable stars. Their periods range from about 10 to about 30 days. They are also known as Type II *Cepheids.

X

X-Ray Astronomy. X-rays are very short electromagnetic radiations, with wavelengths of from 0·1 to 100 *Ångströms. In 1962 United States rocket workers sent up an instrument-carrying vehicle which detected X-rays coming from outer space, and in the following year two intense X-ray sources were found, one in Taurus (the *Crab Nebula) and one in Scorpio. Many others have since been detected, some of which are members of binary systems. They are of immense interest and importance to astronomers, and some of them are very variable.

Y

Year. The time taken for the Earth to go once round the Sun; in round numbers, 365 days. Astronomically, however, there are several different kinds of 'years'.

The *sidereal year* (365·26 days, or 365 days 6 hours 9 minutes 10 seconds) is the true revolution period of the Earth.

The *tropical year* (365·24 days, or 365 days 5 hours 48 minutes 45 seconds) is the time-interval between successive passages of the Sun across the *First Point of Aries. The First Point is not quite stationary; due to *precession it shifts slightly, which is why the tropical year is approximately twenty minutes shorter than the sidereal year.

The *anomalistic year* (365·26 days, or 365 days 6 hours 13 minutes 53 seconds) is the interval between one *perihelion passage and the next. It is slightly longer than the sidereal year, because the position of the Earth's perihelion in its orbit moves by about 11 seconds of arc annually.

The *calendar year* (365·24 days, or 365 days 5 hours 49 minutes 12 seconds) is the mean length of the year according to the *Gregorian calendar.

Yerkes Observatory. Major American observatory in Wisconsin. Its 40-inch refractor remains the largest telescope of its kind.

Z

Z Camelopardalis Variables. Irregular variable stars. Generally they behave in much the same way as *dwarf novæ, but sometimes remain at 'standstills' for some time – weeks or even months.

Zenith. The observer's overhead point (altitude 90 degrees).

Zenith Distance. The angular distance of a celestial body from the observer's *zenith.

Zenithal Hourly Rate. When a *meteor shower is being watched, an observer will probably miss some of the meteors, because the *radiant of the shower will not be directly overhead. A correction is therefore applied in order to obtain the zenithal hourly rate, or Z.H.R. – that is to say, the number of meteors which would have been seen, per hour, under good conditions with the radiant point overhead. In practice, the observed hourly rate will always be less than the Z.H.R.

Zero Gravity. Weightlessness. When a space-craft is in *free fall, and is moving in space without being artificially accelerated or diverted, a man inside will seem to have no weight at all, and will be able to float around in his cabin – as many astronauts have now actually done.

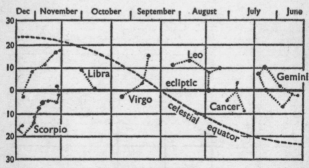

Fig. 98. *The Zodiac*

Zodiac. A belt stretching right round the sky, 8 degrees to either side of the *ecliptic, in which the Sun, Moon, and all principal planets apart from Pluto are always to be found (Fig. 98). It passes through thirteen constellations, the twelve known commonly as the Zodiacal groups plus a small part of Ophiuchus (the Serpent-bearer). The Zodiacal groups are:

> Aries (the Ram),
> Taurus (the Bull),
> Gemini (the Twins),
> Cancer (the Crab),
> Leo (the Lion),
> Virgo (the Virgin),
> Libra (the Scales),
> Scorpio (the Scorpion),
> Sagittarius (the Archer),
> Capricornus (the Sea-Goat),
> Aquarius (the Water-Bearer),
> Pisces (the Fishes).

The Sun, Moon and planets keep near the ecliptic because they move in approximately the same plane; only Mercury has an orbital inclination of more than five degrees. Pluto is the exception. It has an inclination of 17 degrees, so that it may move further away from the ecliptic than the other planets can ever do. However, Pluto is at present in Leo, and it will remain within the Zodiac for the rest of the 20th century.

Zodiacal Light. A cone of light rising from the horizon and

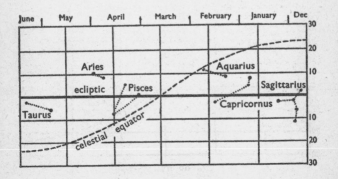

stretching along the *ecliptic (Fig. 99). It is visible only when the Sun is a little way below the horizon, and from Britain it is never very conspicuous, though it may often be observed after sunset in March or before sunrise in September. It is thought to be due to small particles scattered near the main plane of the Solar System. A still fainter extension along the ecliptic is known as the *Zodiacal Band*.

Fig. 99. *Zodiacal light*